Data Science and Machine Learning Applications in Subsurface Engineering

Editor

Daniel Asante Otchere

Pennsylvania State University
University Park, PA, USA

CRC Press is an imprint of the
Taylor & Francis Group, an **informa** business

A SCIENCE PUBLISHERS BOOK

First edition published 2024
by CRC Press
2385 NW Executive Center Drive, Suite 320, Boca Raton FL 33431

and by CRC Press
4 Park Square, Milton Park, Abingdon, Oxon, OX14 4RN

© 2024 Daniel Asante Otchere

CRC Press is an imprint of Taylor & Francis Group, LLC

Reasonable efforts have been made to publish reliable data and information, but the author and publisher cannot assume responsibility for the validity of all materials or the consequences of their use. The authors and publishers have attempted to trace the copyright holders of all material reproduced in this publication and apologize to copyright holders if permission to publish in this form has not been obtained. If any copyright material has not been acknowledged please write and let us know so we may rectify in any future reprint.

Except as permitted under U.S. Copyright Law, no part of this book may be reprinted, reproduced, transmitted, or utilized in any form by any electronic, mechanical, or other means, now known or hereafter invented, including photocopying, microfilming, and recording, or in any information storage or retrieval system, without written permission from the publishers.

For permission to photocopy or use material electronically from this work, access www.copyright.com or contact the Copyright Clearance Center, Inc. (CCC), 222 Rosewood Drive, Danvers, MA 01923, 978-750-8400. For works that are not available on CCC please contact mpkbookspermissions@tandf.co.uk

Trademark notice: Product or corporate names may be trademarks or registered trademarks and are used only for identification and explanation without intent to infringe.

Library of Congress Cataloging-in-Publication Data (applied for)

ISBN: 978-1-032-43364-6 (hbk)
ISBN: 978-1-032-43365-3 (pbk)
ISBN: 978-1-003-36698-0 (ebk)

DOI: 10.1201/9781003366980

Typeset in Times New Roman
by Radiant Productions

Dedication

To God, who has been my rock and guide throughout my life's journey, I dedicate this book with all my heart.

To my dad, Retired Assistant Commissioner of Ghana Police, Daniel Asante Otchere, thank you for instilling in me a strong work ethic and dedication to excellence. You have always been a source of inspiration, and I am grateful for your unwavering support.

To my mum, Jane Duah Otchere, your love and encouragement have been a constant source of strength. Thank you for being my pillar of support and for believing in me even when I doubted myself.

To my beloved, Annabelle, thank you for being my partner, my best friend, and my biggest cheerleader. Your unwavering love and support have been instrumental in my success, and I am grateful for your presence in my life.

To my siblings, Nicholas, Elliot, Yvonne, and Thelma, thank you for your love, support, and unwavering belief in me. You have been a constant source of motivation, and I am grateful for the joy and laughter you bring into my life.

This book is a testament to the love, support, and guidance of these amazing individuals in my life. Thank you all for being part of my life's journey and for being my inspiration.

Foreword

It is with great pleasure that I write the foreword for this book on Data Science and Machine Learning Applications in Subsurface Engineering.

The field of subsurface engineering is a critical aspect of the global energy industry, and it has undergone significant transformations in recent years with the advent of data science and machine learning. This has led to increased efficiency, improved decision-making, and reduced costs in the exploration, production, and development of subsurface resources. Applying data science and machine learning techniques to subsurface engineering is transforming how we approach and solve complex problems in this domain. The combination of vast amounts of data generated by sensors and other sources and the development of powerful algorithms and computing capabilities has enabled us to extract valuable insights and make informed decisions.

The edited book on Data Science and Machine Learning Applications in Subsurface Engineering is an important contribution to this rapidly evolving field. This book brings together a collection of chapters highlighting innovative and impactful applications of these technologies in subsurface engineering. From reservoir characterisation to production optimisation and drilling engineering, the contributors provide various perspectives, demonstrating potentials and challenges. The expert and knowledgeable contributors come from diverse backgrounds, including academia, industry, and government. The diversity of their perspectives enriches the discussion and highlights the need for cross-disciplinary approaches to solving the complex challenges of subsurface engineering.

The chapters in this book cover a wide range of application areas, including data-driven workflows for subsurface characterisation at the well scale and the reservoir scale, machine learning techniques for well performance analysis, and smart completions. Diverse data types are analyzed including well flow data, wireline log data and seismic images. A variety of machine learning algorithms are described, from traditional multivariate statistical methods and tree-based methods to artificial neural networks and deep convolutional neural networks. This book offers valuable insights into the ongoing research and development in the field of subsurface engineering.

Looking forward, we can expect advancements in artificial intelligence and generative machine learning models to improve our ability to analyse subsurface data, extract meaningful insights, and make decisions. However, as with any rapidly evolving field, there will be challenges along the way. It will be essential to continue to develop and refine algorithms to ensure that the models are accurate and reliable and to address the ethical implications of using these technologies in subsurface engineering. I hope this book will serve as a valuable resource for researchers, engineers, and decision-makers in the energy industry and inspire continued innovation and progress in the dynamic field of subsurface engineering.

Tapan Mukerji
Professor (Research)
Energy Science & Engineering
Stanford University

Preface

Data science and machine learning have revolutionised various industries, and the subsurface engineering industry is no exception. Subsurface engineering involves the exploration, development, and production of natural resources like oil and gas, and is critical to meet the world's energy demands. The application of data science and machine learning techniques in subsurface engineering has the potential to significantly improve the efficiency, accuracy, and safety of operations in this industry.

This research book on Data Science and Machine Learning Applications in Subsurface Engineering provides a comprehensive overview of this field's latest research and developments. The book is intended for professionals, researchers, and students interested in understanding the potential of data science and machine learning in subsurface engineering. I hope that readers value each chapter in this book because subsurface engineering plays a vital role in meeting the world's energy demands, and the application of data science and machine learning techniques has the potential to significantly improve efficiency and accuracy, and safety of operations in this industry.

This book brings together a collection of chapters that showcase some of the most innovative and impactful applications of these technologies in subsurface engineering. The contributors to this book represent a diverse range of backgrounds and expertise, from academic researchers to industry professionals. Each chapter offers a unique perspective on using data science and machine learning in subsurface engineering. The success of this book is a testament to our contributors' hard work and dedication. I extend my deepest gratitude to Ramez, Daniel, Eric Thompson, Ayoub, Halim, and Nikita for their contributions. The contributions of our authors have made this book possible, and their research provides valuable insights into the potential of data science and machine learning in subsurface engineering. I want to sincerely thank each of our contributors for their hard work, dedication, and commitment to advancing knowledge in this field.

Throughout the book, several key themes emerge, including the importance of interdisciplinary collaboration, the need for robust and transparent data management practices, and the challenges of implementing data science and machine learning techniques in real-world environments. These themes reflect the complex and evolving nature of subsurface engineering and highlight the need for continued innovation and collaboration in this field.

One of the most significant challenges in this field is the integration of data science and machine learning techniques with existing engineering practices. The contributors to this book demonstrate that successful integration requires a deep understanding of both the technical and practical aspects of subsurface engineering. This research book contributes to the advancement of the field by providing a platform for collaboration and sharing knowledge, expertise, and experiences in applying data science and machine learning in subsurface engineering.

Another key theme that emerges from this book is the importance of transparency and ethical considerations in the use of data science and machine learning in subsurface engineering. The contributors to this book emphasise the need for transparency in data collection, analysis, and interpretation, as well as the importance of considering the potential impacts of these techniques on the environment and society. It is known that many articles and texts prevent the reader from reproducing the results either because the data were not freely available or because the software was inaccessible, or only available for purchase. Therefore, it was my goal to be as hands-on as possible in stating the methods used, enabling the readers to reproduce the results and extend the methodology to their own data. Furthermore, the authors opted to use the Python language, an open-access software, for all stages of the machine learning process.

Most of the chapters contain links to the data sets, Python notebooks, and software used here to reproduce the analyses in each chapter. I selected Python as the computational engine of this book for several reasons. First, Python is freely available for multiple operating systems. I encourage you to take the time to compare each of your solutions with the results in this book. Performing this comparison may help you become more familiar with a technique you could have used to solve a specific problem more efficiently. In some cases, it could also reveal that you have discovered a better or simpler way to solve the problem than the chapters had.

This research book is an important contribution to the field. The application of data science and machine learning techniques in subsurface engineering has the potential to significantly improve the efficiency, accuracy, and safety of operations in this industry. The research presented in this book showcases novel techniques and applications that can address some of the most pressing challenges in subsurface engineering, including seismic interpretation, reservoir characterisation, production optimisation, and data-driven decision-making. The research offers a unique perspective on the current state-of-the-art and future directions in this field, making it a valuable resource for researchers, practitioners, and students in this domain.

This edited research book on Data Science and Machine Learning Applications in Subsurface Engineering features a collection of research papers highlighting the latest field advances. The book covers a wide range of topics related to the application of data science and machine learning techniques in subsurface engineering. The book's main contributions can be summed as novel techniques for seismic interpretation, improved reservoir characterisation, data-driven decision-making, and applications beyond oil and gas. Overall, the contributions of this edited research book advance our understanding of the potential of data science and machine learning in subsurface engineering. The book provides a valuable resource for researchers, practitioners, and students interested in this field and highlights the latest developments that can be applied to solve complex problems in subsurface engineering.

The future of subsurface engineering is exciting and filled with opportunities, but it also presents significant challenges. The rapid pace of technological innovation and the increasing demand for sustainable energy solutions require the industry's continuous and dynamic response. The contributors to this book provide valuable insights into the potential of data science and machine learning techniques to address these challenges, as well as the need for continued collaboration and innovation in this field.

In conclusion, this book offers a comprehensive collection of research papers covering various aspects of applying data science and machine learning in subsurface engineering. The book highlights the potential of these technologies to transform how we approach subsurface engineering challenges and highlights the importance of collaboration, transparency, and ethical considerations in their use. As the editor, I encourage readers to delve into the chapters and explore the exciting research, as they can expect to gain insight into the latest techniques and developments in this exciting field.

Contents

Dedication	iii
Foreword	iv
Preface	vi
1. Introduction	1
2. Enhancing Drilling Fluid Lost-circulation Prediction: Using Model Agnostic and Supervised Machine Learning	6
1. Introduction	6
2. Background of Machine Learning Regression Models	9
3. Data Collection and Description	11
4. Methodology	11
4.1 Data Analysis and Visualisation	11
4.2 Machine Learning Model Application	13
4.3 Explainable AI	15
4.3.1 Permutation Feature Importance	15
4.3.2 Shapley Values	16
5. Results and Discussion	17
5.1 Evaluation of Model Performance	17
5.2 Model Agnostic Results	20
5.3 Analysis of Features Using Model Agnostic Metrics	22
5.4 Analysis of Features Using Shapley Values Model Agnostic Metrics	23
5.5 Evaluation of Top Features	26
5.6 Model Optimisation	26
5.7 Sensitivity Analysis	27
6. Conclusions	28
Acknowledgement	29
Data Availability	29
References	30

3. Application of a Novel Stacked Ensemble Model in Predicting Total Porosity and Free Fluid Index via Wireline and NMR Logs   33
  1. Introduction   33
  2. Nuclear Magnetic Resonance   36
    2.1 Concept and Application   36
    2.2 Works Related to the Use of Machine Learning in NMR for Reservoir Characterisation   38
  3. Methodology   40
    3.1 Data Collection and Description   40
    3.2 Data Analysis and Feature Engineering   42
    3.3 Machine Learning Model Application   43
        3.3.1 Building Deep Learning Models   43
        3.3.2 Building a Hybrid Stacked Ensemble Model   44
    3.4 Criteria for Model Evaluation   46
  4. Results and Discussion   47
    4.1 Evaluation of Models' Performances   47
  5. Conclusions   54
  Acknowledgement   55
  References   55

4. Compressional and Shear Sonic Log Determination: Using Data-Driven Machine Learning Techniques   57
  1. Introduction   57
  2. Literature Review   58
  3. Background of Machine Learning Regression Models   64
    3.1 Decision Tree Conceptual Overview   64
        3.1.1 Attribute Selection Measures   65
    3.2 Random Forest Conceptual Overview   68
    3.3 Extremely Randomised Trees Conceptual Overview   69
  4. Data Collection and Description   70
  5. Methodology   73
    5.1 Data Analysis and Visualisation   73
    5.2 Machine Learning Model Application   73
  6. Results and Discussion   75
    6.1 Evaluation of Model Performance   75
    6.2 Model Optimisation   79
    6.3 Model Deployment   82
  7. Conclusions   82
  Acknowledgement   83
  Data Availability   83
  References   83

5. Data-Driven Virtual Flow Metering Systems — 87
1. Introduction — 87
2. VFM Key Characteristics — 88
3. Data Driven VFM Main Application Areas — 90
   - 3.1 Virtual Sensing in ESP Wells — 90
   - 3.2 Virtual Sensing for SRP Wells — 91
     - 3.2.1 Virtual Flow Meter on Rod Pumping Systems — 93
     - 3.2.2 Virtual Sensing of the Dynamometer Card — 93
   - 3.3 Virtual Sensing for Gas Lifted Wells — 94
   - 3.4 Virtual Sensing for Gas Wells and Plunger Lifted Wells — 95
   - 3.5 Miscellaneous Applications for Identifying Flow Regimes — 96
4. Methodology of Building Data-driven VFMs — 97
   - 4.1 Data Collection and Preprocessing — 97
   - 4.2 Model Development — 98
5. Field Experience with a Data-driven VFM System — 99
References — 100

6. Data-driven and Machine Learning Approach in Estimating Multi-zonal ICV Water Injection Rates in a Smart Well Completion — 104
1. Introduction — 104
2. Brief Overview of Intelligent Well Completion — 107
   - 2.1 ICV Setting and Determination — 107
   - 2.2 Literature Review of ICV Innovations and Machine Learning Applications — 108
3. Methodology — 110
   - 3.1 A Brief Overview of Models Used in This Study — 111
   - 3.2 Criteria for Model Evaluation — 112
4. Results and Discussion — 114
   - 4.1 Explainable AI — 114
   - 4.2 Model Evaluation — 116
   - 4.3 Sensitivity Analysis — 117
   - 4.4 Model Deployment — 120
5. Conclusions — 122
Code Availability — 122
Acknowledgement — 123
References — 123

7. Carbon Dioxide Low Salinity Water Alternating Gas (CO$_2$ LSWAG) Oil Recovery Factor Prediction in Carbonate Reservoir: Using Supervised Machine Learning Models — 125
1. Introduction — 125
2. Methodology — 132

  2.1 Modeling of CO_2-LSWAG 132
  2.2 Geochemical Reactions of CO_2-LSWAG 133
  2.3 Machine Learning Methods 133
    2.3.1 Multivariate Adaptive Regression Splines (MARS) 134
    2.3.2 Group Method of Data Handling (GMDH) 136
    2.3.3 Performance Metrics 137
    2.3.4 Dataset Standardisation 138
 3. Results and Discussion 138
  3.1 Numerical Model Description 138
  3.2 Input and Target Dataset 141
  3.3 MARS Modeling 143
  3.4 GMDH Modeling 148
  3.5 Numerical Simulator and Machine Learning 153
    Computational Time
 4. Conclusion 153
 Acknowledgment 154
 References 154

**8. Improving Seismic Salt Mapping through Transfer Learning 159
Using A Pre-trained Deep Convolutional Neural Network: A Case
Study on Groningen Field**
 1. Introduction 159
 2. Method 164
  2.1 Collection and Description of Data 164
  2.2 Deep Convolutional Neural Network in Salt Mapping 164
    and Post-processing
    2.2.1 Simplified Architecture of Residual U-net 168
  2.3 Transfer Learning Application 169
  2.4 Criteria for Model Evaluation 169
 3. Results and Discussion 172
  3.1 Calculated Salt Body Volume 172
  3.2 Semantic Segmentation – Transfer Learning Application 174
  3.3 Sensitivity Analysis of Model and Expert Interpretations 177
 4. Conclusions 178
 Data and Software Availability 179
 Acknowledgement 179
 References 179

**9. Super-Vertical-Resolution Reconstruction of Seismic Volume 181
Using A Pre-trained Deep Convolutional Neural Network:
A Case Study on Opunake Field**
 1. Introduction 181
 2. Brief Overview 183

3. Methodology 187
      3.1 Regional Geological Overview of the Opunake Field 187
      3.2 Local Geological Overview of the Opunake Field 188
      3.3 Deep Convolutional Neural Network in Seismic 189
          Image Resolution
          3.3.1 Simplified Architecture of Residual U-net 190
      3.4 Training and Testing Process 192
      3.5 Criteria for Model Evaluation 193
   4. Results and Discussion 194
      4.1 Conditioned Seismic Volume 194
      4.2 Model Evaluation 197
   5. Conclusions 202
   Data and Software Availability 203
   Acknowledgement 203
   References 203

10. Petroleum Reservoir Characterisation: A Review from Empirical to Computer-Based Applications 207
   1. Introduction 207
   2. Empirical Models for Petrophysical Property Prediction 209
      2.1 Porosity and Permeability Prediction Models 209
      2.2 Saturation Prediction Models 210
   3. Fractal Analysis in Reservoir Characterisation 213
   4. Application of Artificial Intelligence in Petrophysical 214
      Property Prediction
      4.1 Artificial Neural Networks (ANNs) 215
          4.1.1 ANN Application in Petrophysical 218
                Reservoir Prediction
      4.2 Support Vector Machine (SVM) 228
          4.2.1 Machine Learning (ML) Application in 229
                Petrophysical Reservoir Prediction
   5. Lithology and Facies Analysis 234
      5.1 AI Applications in Lithology and Facies Analysis 234
   6. Seismic Guided Petrophysical Property Prediction 239
   7. Hybrid Models of AI for Petrophysical Property Prediction 244
   8. Summary 247
   9. Challenges and Perspectives 248
      9.1 AI Perspective 248
      9.2 Rock Physics Perspective 250
   10. Conclusions 251
   References 253

11. Artificial Lift Design for Future Inflow and Outflow Performance for Jubilee Oilfield: Using Historical Production Data and Artificial Neural Network Models   261

   1. Introduction   261
   2. Methodology   263
      2.1   Artificial Lift Screening Techniques   263
      2.2   Inflow Performance Relationship Production Forecast   263
      2.3   Outflow Performance Relationship Production Forecast   264
      2.4   PROSPER Procedure for Well Model Set-Up   264
          2.4.1   Deviation Survey Data Input   266
          2.4.2   Surface Equipment Data Input   266
          2.4.3   Downhole Equipment Data Input   266
          2.4.4   Average Heat Capacities Data Input   267
      2.5   Artificial Neural Networks   267
          2.5.1   Back Propagation Neural Network   268
          2.5.2   Radial Basis Function Neural Network   268
          2.5.3   ANN Procedure   268
   3. Results and Discussion   269
      3.1   Production and Well Data of the Study Area   269
          3.1.1   Base Case Flow Rates   271
      3.2   Artificial Lift Screening   272
      3.3   PROSPER Simulation Results   273
          3.3.1   IPR Curves   273
          3.3.2   Vertical Lift Performance Correlations   275
          3.3.3   Desired Flow Rates   276
      3.4   Gas Lift Results   277
          3.4.1   Optimum Production Rates   278
      3.5   ANN Results   279
          3.5.1   ANN Architecture   279
          3.5.2   Model Visualization   280
      3.6   Discussion   282
   4. Conclusions   283
   Acknowledgment   284
   References   284

12. Modelling Two-phase Flow Parameters Utilizing Machine-learning Methodology   286

   1. Introduction   286
   2. Data Sources and Existing Correlations   288
   3. Methodology   289

4.	Results and Discussions	291
	4.1 Data Pre-processing	291
	4.2 Model Development and Evaluation	293
5.	Comparison between ML Algorithms and Existing Correlations	295
6.	Conclusions and Recommendations	298
	Nomenclature	298
	References	299

Index   **303**

CHAPTER 1
Introduction

"Data is the new oil"—this buzz phrase has become a ubiquitous adage in today's digital age. Data science and machine learning have become indispensable tools for extracting insights and value from vast amounts of data. The field of subsurface engineering is no exception to this trend, and this book brings together a collection of chapters from experts in this field, exploring various data science and machine learning applications in subsurface engineering.

In this book, we delve into the potential of these technologies for subsurface engineering, including topics such as data-driven reservoir characterisation, machine learning in drilling operations, and computer vision application in seismic image processing and interpretation. The book is divided into several chapters, each offering a unique perspective and case studies on the different applications of data science and machine learning in subsurface engineering.

Chapter 2 focuses on predicting drilling fluid loss in the Marun oil field using machine learning models. The research begins by assessing the importance of input features through model agnostic metrics. Once a suitable dataset is established, several machine learning models will be used, with the best-performing one optimised using the Bayesian Optimisation algorithm. This research aims to generate new insights into the generalisation of individual features to explain the target. The study's main contributions are twofold. Firstly, it provides a global explanation of variables in mud-loss prediction using explainable artificial intelligence (AI). Secondly, it develops a machine learning workflow that utilises explainable AI to enhance drilling fluid lost circulation prediction. Overall, this research offers valuable insights into applying machine learning to subsurface engineering, and the potential benefits of explainable AI in improving drilling fluid loss prediction.

Chapter 3 describes a study aimed at developing an AI-based model that can predict porosity and producible pore volume fraction using wireline logs and NMR-measured total porosity and free fluid index. Wireline logs are more economical and time-saving than running nuclear magnetic resonance (NMR) logs in all wells, making it a desirable field-scale technique for quantitative porosity and pore volume fraction measurement. The study aims to provide a more efficient and accurate method for reservoir characterisation and management, especially given the current demand for inexpensive and reliable techniques in the face of high oil prices. By developing an accurate prediction model, the study hopes to assist in improving reservoir management and decision-making processes.

Chapter 4 delves into the use of data-driven machine learning techniques to determine compressional and shear sonic logs, which are crucial in subsurface engineering. The authors discuss the importance of using relevant input variables to train supervised machine learning models, as the model's performance can degrade if irrelevant data or essential features are omitted. The authors identify and select input features based on their demonstrated impact on Vs and Vp measurements to address this issue. The study is distinguished from earlier research by its multi-output prediction of Vp and Vs, with Vp not being used as an input for Vs prediction. The authors use data from three wells in the Volve oil field to evaluate the relevance of input features, and different machine learning models are employed to predict Vp and Vs. The authors then use statistical metrics to determine the best-performing model, which is optimised using the Bayesian Optimisation (BO) algorithm.

Chapter 5 explores the concept of virtual flow metering (VFM) in the oil and gas industry and focuses on data-driven solutions. VFM technology relies on analytical or data-driven models for real-time calculations of phase production. The chapter first introduces the classification of VFM based on modelling paradigms, including the first principle and data-driven VFMs. The applications of data-driven modelling in VFM systems are then discussed, emphasising the models' features, predicted variables, input data, and respective papers. The chapter also describes the components and methodology used to develop data-driven virtual flow meters using the cross-industry standard process for a data mining framework. An implementation of this methodology is presented to estimate flow rate and water cut prediction from the electrical submersible pump's sensor data. Finally, the chapter discusses the real operational experience reported in the literature using data-driven models. Overall, this chapter provides a comprehensive overview of the concept of VFM and highlights the potential applications of data-driven solutions in VFM systems in the oil and gas industry.

Chapter 6 presents a novel approach to optimising inflow control valve (ICV) settings to increase oil production using machine learning techniques. The chapter proposes a data-driven approach to estimate the injected water volumes into each reservoir when ICV settings formula is discordant rather than recalibrating empirical correlation, which leads to interruptions in field production. The study employs eight machine learning models to estimate each reservoir unit's volume of production fluids based on operational parameters, providing a simple and accurate approach to reservoir management plans. The method can also provide real-time estimates of produced volumes, allowing daily production operational changes to meet critical targets and develop domestic oil resources responsibly. This initiative has tremendous research possibilities, and the proposed approach could be applied to oil production wells with various input parameters upon acceptable results.

Chapter 7 of this book explores using Carbon Dioxide Low Salinity Water Alternating Gas (CO_2 LSWAG) flooding as an Enhanced Oil Recovery (EOR) technique in carbonate reservoirs. The chapter highlights the benefits of this technique, including a high recovery factor and improved displacement efficiencies. The authors use a compositional simulator with geochemical models to develop proxy models for predicting the oil recovery factor. Multivariate Adaptive Regression Splines (MARS) and Group Method of Data Handling (GMDH) machine learning methods are used in the study to develop these proxy models. The authors advocate for using machine learning proxy models as prediction tools to enhance the efficient full-field implementation of this technique and reduce the computational time associated with numerical simulations in carbonate reservoirs.

Chapter 8 showcases the application of transfer learning to a convolutional neural network pre-trained with synthetic labels, generating salt probability models for use in seismic imaging and velocity modelling phases. The use of deep learning techniques in object and edge detection has succeeded in various fields, making them a promising approach for seismic salt mapping. The study's main contributions are improved accuracy in salt segmentation, time and cost savings, enhanced data analysis, and potential for future research. By using transfer learning to automate the process of salt segmentation in seismic images, we can save time and reduce costs while improving accuracy, ultimately improving our understanding of the subsurface geology and advancing the energy transition journey. The findings of this study will provide new insights into the use of transfer learning for salt segmentation in seismic images, highlighting the potential for developing new energy resources, and mitigating climate change.

Chapter 9 explores advanced AI techniques to improve seismic image resolution and quality, which is crucial for accurate subsurface exploration and analysis. This study focuses on using a pre-trained deep convolutional neural network (DCNN) to enhance the signal-to-noise ratio (SNR) and vertical resolution in seismic images of the Opunake field. The study compares two-image resolution DCNN techniques to identify the most suitable one for vertical resolution improvements. The success of using synthetic data to pre-train the DCNN and deploying it on a real field makes AI a promising candidate for enhancing vertical and lateral seismic image resolution. This research highlights the potential of AI techniques in improving seismic imaging and interpretation, paving the way for future developments in this area.

Finally, Chapter 10 provides a structured overview of the application of AI in reservoir characterisation. The paper focuses on the empirical approach, AI application, and its enhancement with other computational models, as well as recent AI advancements for reservoir property prediction. The paper is divided into 10 sections, with Section 2 providing an overview of the empirical correlation used for petrophysical reservoir characterisation. The subsequent sections delve into the application of fractals analysis, AI, and machine learning techniques such as ANN and SVM and the review of lithology and facies analysis. The contribution of seismic data in reservoir property prediction is also examined, with Section 7 exploring the application of hybrid AI in reservoir characterisation. Section 8 highlights the summary of individual AI algorithms used in reservoir characterisation, while Section 9 addresses the challenges and the way forward for reservoir characterisation. Finally, Section 10 concludes the study and spells out the significant keys drawn from the present review. Overall, this structured overview provides valuable insights into the application of AI in reservoir characterisation, highlighting its potential to enhance the accuracy and efficiency of characterisation processes.

Together, these chapters provide a comprehensive overview of how data science and machine learning can be applied to subsurface engineering, demonstrating the potential of these technologies for revolutionising the way we explore, produce, and manage subsurface resources.

Subsurface engineering is a complex and challenging field that requires a deep understanding of the physical properties of the subsurface, including geology, fluid mechanics, and rock mechanics. Traditionally, subsurface engineering relied on deterministic models and manual interpretation of data, which were often time-consuming and prone to errors. The advent of data science and machine learning has transformed this field, offering new tools and approaches for handling and analysing large volumes of subsurface data. These technologies have revolutionised subsurface engineering by enabling more accurate predictions, faster decision-making, and improved resource management. This book provides a timely and comprehensive overview of the different applications of data science and machine learning in subsurface engineering, highlighting the potential of these technologies for unlocking new insights and value from subsurface data. By exploring the different topics covered in this book, readers will gain a deeper understanding of the transformative power of data science and machine learning in subsurface engineering and the challenges and opportunities of integrating these technologies into subsurface workflows.

In conclusion, this book offers a comprehensive overview of the different applications of data science and machine learning in subsurface engineering, demonstrating the potential of these technologies for unlocking new insights and value from subsurface data. The book's central themes include the importance of interdisciplinary collaboration, the need for robust and transparent data management practices, and the challenges of implementing data science and machine learning techniques in real-world environments. In addition, the book highlights the importance of transparency and ethical considerations in using data science and machine learning in subsurface engineering, emphasising the need for responsible and ethical practices in this rapidly evolving field.

By bringing together experts from various fields, this book offers a unique and interdisciplinary perspective on applying data science and machine learning in subsurface engineering, highlighting the potential of these technologies for transforming the way we explore, produce, and manage subsurface resources. The book's contributions to the field include case studies, best practices, and critical analyses of the opportunities and challenges of implementing data science and machine learning techniques in subsurface engineering. Overall, this book serves as a valuable resource for researchers, practitioners, and students in the field of subsurface engineering, offering insights and perspectives that are critical for staying up-to-date with the latest developments in this rapidly evolving field.

CHAPTER 2

Enhancing Drilling Fluid Lost-circulation Prediction

Using Model Agnostic and Supervised Machine Learning

Daniel Asante Otchere,[1,2,*]
Mohammed Ayoub Abdalla Mohammed,[3] *Hamoud Al-Hadrami*[4]
and *Thomas Boahen Boakye*[5]

1. Introduction

Drilling is a complex and high-risk element of oil and gas field development and production (Zarrouk and McLean, 2019). Drilling fluid serves several purposes in rotary drilling, where the drilling mud is cycled through the drill string to remove cuttings and improve drill bit performance (Alkinani et al., 2020). The size, shape, and density of the drilling fluid's cuttings, as well as its annular velocity, all affect the drilling fluid's ability to remove accumulated cuttings. One of the most prevalent drilling industry difficulties is drilling fluid lost returns. This problem occurs whenever the volume of mud injected during drilling partially or totally filtrates into the formation rather than flowing back up to the surface (Toreifi et al., 2014). Due to the volume

---
[1] Centre of Research for Subsurface Seismic Imaging, Universiti Teknologi PETRONAS, 32610, Seri Iskandar, Perak Darul Ridzuan, Malaysia.
[2] Institute for Computational and Data Sciences, Pennsylvania State University, University Park, PA, USA.
[3] Chemical and Petroleum Engineering, UAE University, Sheik Khalifa Street at Tawam R/A, Maqam District, Al Ain, United Arab Emirates.
[4] Department of Petroleum and Chemical Engineering, Sultan Qaboos University, Muscat, Oman.
[5] Subsurface Department, Tullow Ghana Limited, North Dzorwuku, Off George Bush Highway, PMB, Accra, Ghana.
* Corresponding author: ascotjnr@yahoo.com

of mud lost during drilling, Pilehvari and Nyshadham (2002) proposed three classifications due to the severity or otherwise. The three classifications are defined as seepage (ranging from 0.5 to 10 bbls/hr), partial (ranging from 10 to 500 bbls/hr), and complete (above 500 bbls/hr) losses of all the fluid. The losses can be attributed to the formation type and structural and petrophysical properties of the formation (Moazzeni et al., 2012). The loss of these drilling fluids leads to increased costs in drilling operations, differential sticking, a blowout, damage to reservoir intervals, and, most seriously leading to a loss of the well (Alkinani et al., 2019a; Feng and Gray, 2017; Sabah et al., 2019; Toreifi et al., 2014).

Minimising the loss of drilling fluids is of the utmost importance considering the severe consequences. This fluid loss has led to the development of different preventive and remedial treatments employed to prevent fluid loss. Before deciding on the appropriate lost circulation solution to utilise to reduce mud loss, the degree of the mud loss should be determined. It should, however, be noted that finding a single solution to reduce lost circulation is challenging. Consequently, several lost circulation remedial alternatives are available, including but not limited to fibrous, high viscosity pills, granular, brittle materials, and cement sludges. Each treatment method or material is chosen depending on how much of the mud is lost, the time and expense involved, the drilling phase, the fluid type, and the theft zone (Alkinani et al., 2020). The main reason why these treatments are used is to bridge existing fractures and prevent the development of new fractures (Alkinani et al., 2019b).

Several factors lead to this drilling problem, the most severe being formation, drilling operation, and time-dependent. Since these factors are highly dimensional and have complex relationships, traditional mathematical techniques have proven futile in predicting drilling fluid loss (Sabah et al., 2019; Toreifi et al., 2014). As a result, a significant amount of effort and money is spent harnessing technical breakthroughs in data acquisition to potentially limit its likelihood. One method for accomplishing this is accurately forecasting lost circulation using drilling parameters. Artificial Intelligence (AI) techniques are now widely employed in the oil and gas sector, resulting in substantial advancements in their use to increase accuracy (Otchere et al., 2021a). Several studies have also employed machine learning models to forecast drilling fluid loss circulation, demonstrating their efficacy in creating patterns among complicated drilling parametric interactions (Alkinani et al., 2020; Sabah et al., 2019; Toreifi et al., 2014).

Sabah et al. (2021) introduced a hybrid machine learning approach that enhances the prediction of lost circulation using the Multilayer Perceptron Neural Network (MLP-NN), and the Least Squares Support Vector Machine (LSSVM) models. To enhance the accuracy of the predictions, feature selection was employed, which helped to eliminate input parameters that were not beneficial to the machine learning model. The researchers applied a Savitzky–Golay (SG) filter to reduce noise recorded in the data. From their results, the wrapper method, the most efficient feature selection technique, was used in conjunction with several evolutionary algorithms to improve model performances. Their study concluded that the LSSVM-Cuckoo Optimisation Algorithm recorded the highest prediction accuracy of 0.94 compared to all the hybrid models developed. Alkinani et al. (2020) also employed an artificial neural network to estimate lost circulation in induced and natural fractures. Their work entailed separating the datasets into three categories: 60% for training, 20% for testing, and 20% for validating. Databases, consisting of nine input features for each fracture type, were created for both natural and induced fractures to give room for data normalisation. The application of feature selection justified the selection of the Levenberg–Marquardt function to train the dataset as it resulted in the highest accuracy. Their results achieved an accuracy of 0.96 and 0.93, respectively, for both natural and induced fracture networks.

Abbas et al. (2019b) assessed the potential of machine learning in mining and analysing drilling data to predict lost circulation while drilling. For the selection of input parameters, the work made use of the feature ranking method to reduce the dimensionality of the dataset. Eighteen out of the 23 studied parameters were selected as input variables to predict the lost–circulation. The feature ranking method enhanced the efficiency of the dataset, thereby improving computational processing time. The results showed that the Gaussian kernel support vector machine (SVM) recorded the highest accuracy of 0.92. Their work led to more research, such as using two different machine learning models that consider geological and operational parameters in making lost-circulation predictions (Abbas et al., 2019a). Toreifi et al. (2014) also proposed a new technique using the modular neural network (MNN) and particle swarm optimisation (PSO) capable of predicting drilling fluid loss. Data normalisation was used to transform the data into a range between 0 and +1 to develop a more efficient model, for which 60% was used for training, 20% for testing, and 20% for validating. Their results concluded that the PSO achieved a more accurate output by optimising the parameter variation process. Other influential research and review work in the field of integrating AI with drilling activities include Aalizad and Rashidinejad (2012), Abbas et al. (2019c), Ahmadi (2016), Al-Baiyat and Heinze (2012), Barbosa et al. (2019), Brankovic et al. (2021).

Considering different machine learning models and techniques used in predicting drilling fluid loss, the main difference was the type of model being used and the input variables. Supervised machine learning models are only as efficient as the information they are trained with. Therefore, when irrelevant data is included as input, the model's performance suffers (Otchere et al., 2021c). As such, there is the need to develop a workflow capable of determining the most appropriate features, which involves engineering new features from existing input variables (Otchere et al., 2021b). These new features should be capable of improving model performance compared to when they are inputted singularly. Again, the choice of machine learning models should not be a hindrance if a more robust model appending all robust models into one super machine learning model is used. This approach tends to mitigate the inherent problems of each model while complementing each other's strengths.

This study uses data from the Marun oil field used by Sabah et al. (2021) and Toreifi et al. (2014). The input features and their importance will be assessed using model agnostic metrics when predicting drilling fluid loss. Once a suitable dataset of relevant features is established, several machine learning models will be used to predict drilling fluid loss. The model that gives the best results will then be optimised using the Bayesian Optimisation (BO) algorithm. The main objective of this research is to analyse the effect of individual features to generate new insight into their generalisation of explaining the target. The main contributions of this research are the global explanation of variables in mud-loss prediction using an explainable AI and the development of a machine learning workflow to enhance the prediction of drilling fluid lost circulation using an explainable AI.

2. Background of Machine Learning Regression Models

This research stems from earlier studies where several workflows and techniques were used to predict drilling fluid loss using available input variables. As academics increasingly move away from empirical correlations, the use of machine learning has become entrenched. Individual drilling parameters offer critical information concerning drilling fluid lost circulation. However, when some subsets are used, they can predict drilling fluid loss more accurately. In some instances, researchers employ the Pearson correlation to identify appropriate variables, while others use wrapper approaches, intrinsic techniques, and metaheuristics algorithms to identify important features. The use of several techniques indicates that no definitive collection of features exists for making this prediction. However, a robust workflow is needed to utilise model agnostic techniques to solve this prediction problem. This study aims to understand the relationship among key input variables relevant for drilling fluid lost-circulation prediction. The relevant features selected by the model agnostic techniques will then be used as input features. Several machine learning models will be used to predict fluid loss based on their learning theory and ability to work with high-dimensional and complex data.

The reviewed models perform differently based on learning theory, dimensionality, small or large data, and different data distribution. Identifying a suitable model that can handle all these problems has become necessary since drilling data are commonly highly dimensional, either small or large, with different data distributions. Based on these assertions, this study establishes relevant features that can lead to lower prediction errors. Improving the accuracy in estimating mud loss has a massive impact on drilling operations and the integrity of a well. Hence, the application of machine learning models, although dependent on data type, can be applied to drilling operations with similar input features. The most common issue with machine learning models and their varying performance is centred on data. Having the capability of explaining the input variables and determining causation has been the pinnacle of this study as new approaches are created to solve this issue. The slightest gain in accuracy is critical in improving decision-making in the petroleum business, making this field of research critical. Table 2.1 presents a summary analysis of the models reviewed in this research.

Table 2.1. Summary of algorithms used in this study and corresponding authors (Otchere et al., 2022b).

Model	Description	Developed or Implemented by
Ridge Regression	Robust technique against multicollinearity minimises standard errors by applying some bias to the model estimates, resulting in a more reliable prediction. Ridge regression model works by calculating the difference between the means of the standardised dependent and independent variables and dividing by their standard deviations.	(Hoerl and Kennard, 1970)
Least Absolute Shrinkage and Selection Operator (Lasso)	Robust against multicollinearity, perform better with lower-dimensional data, prevents overfitting, and reduces standard errors.	(Tibshirani, 1996)
Support vector machine (SVM)	Utilises the Structural Risk Minimisation induction principle, hence, produces improved generalised global, sparse, and unique results with a simple geometric interpretation.	(Vapnik and Lerner, 1963)
Decision Tree (DT)	Uses induction and pruning techniques to build hierarchical decision boundaries and remove unnecessary structures from the decision tree to battle overfitting.	(Gordon et al., 1984)
Extreme Gradient Boosting (XGBoost)	An implementation of gradient boosted decision trees of the first- and second-order to maximise the loss function, adding an extra regularisation term to avoid overfitting by adjusting the final weights.	(Chen and Guestrin, 2016)
Extremely Randomised Trees (Extra Tree or ET)	Potentially achieves better performance than the random forest, a simpler algorithm is employed to construct the decision trees used as ensemble members.	(Geurts et al., 2006)
Random Forest (RF)	An advanced decision tree that is robust against overfitting and offers easy interpretability.	(Breiman, 2001)

3. Data Collection and Description

This research aims to improve the prediction of drilling fluid loss using data from the onshore Marun oilfield during drilling operations. Fracturing in the reservoir interval is a defining feature of this field in Iran's Khuzestan province. The data consists of 2,820 data points with 19 features compiled from resources such as final well reports, daily drilling reports (DDRs), literature, and daily mud reports (Sabah et al., 2021; Toreifi et al., 2014). However, after removing the missing data, 2783 points remained. Table 2.2 shows the descriptive statistics of some selected features. The main input features for this research are;

1. Drilling operation features: Geographic coordinates, depth where the loss occurred, weight on bit (WOB), hole size, drilling time, penetration rate (RPM), mud-pump pressure, the flow rate of the mud pump, Marsh funnel viscosity (MFVIS), and bit rotational speed.
2. Characteristics of drilling mud: Viscosity, shear stresses at shear rates of 600/300 rpm, Gel strength (10-s and 10-min), retort solid per cent, and mud weight (MW).
3. Formation characteristics: Formation type, pore pressure, and fracture pressure.

One of the main advantages of applying data analytics and machine learning to data is to find patterns and hidden information in high-dimensional data. As such, some input variables will be deemed irrelevant in predicting drilling fluid lost circulation. For this study, the lithologies were encoded into numeric variables ranging from 1–15 (Table 2.3). The coordinates for each well were also removed.

4. Methodology

4.1 Data Analysis and Visualisation

Data analysis and visualisation were used to aid in understanding how the input features related to the output. The degree of correlation among each input variable pair and output was quantified using the Spearman rho covariance matrix. From the heatmap, the pump pressure was the only feature to show a moderate correlation to the target. The pair plot also depicted the nonlinear distribution between some of the input and the target. Based on the nonlinear distribution depicted in Fig. 2.1, it is evident that none of the features showed a linear correlation to the target. Hence, nonlinear models were deemed appropriate for this research.

Table 2.2. Summary statistics of some input and target variables.

	Hole Size	WOB (1000lb)	Flow Rate (gpm)	MW (pcf)	MFVIS	Retort Solid (%)	Pore Pressure (psi)	Fracture Pressure (psi)	Fan 600/Fan 300	Gel10min/Gel10s	Pump Pressure (psi)	RPM	Mud Loss (bbl/hr)
count	2783	2783	2783	2783	2783	2783	2783	2783	2783	2783	2783	2783	2783
mean	12.3	20.9	545.6	95.3	47	23	5421.4	7929.8	1.6	1.5	1934.4	137.9	56.6
std	4.9	9.4	275.1	34.6	11.9	16.6	2508.2	2760.4	0.2	0.2	721.7	34.6	97.8
min	4.1	1	80	30.5	27	0	1254.8	1970.6	1.2	1.1	297	42.8	0
25%	8.4	15	280	66	38	8	3470	5468.9	1.5	1.3	1423.4	111.6	0
50%	12.3	20	520	79.3	44	18	4473.4	8736.7	1.6	1.5	2228.5	146.7	17.7
75%	17.5	25	830	140	56	42	7554.4	9687.2	1.8	1.6	2472.3	166.7	68.4
max	26	70	1000	160.5	100	61	10507.9	13610.4	2	2.9	2954.2	198.1	696.9

Table 2.3. Assigned numbers for the Marun Field Formations (Sabah et al., 2021).

Formation Type	Numeric Code
Aghajery	1
Mishan	2
Gachsaran 7	3
Gachsaran 6	4
Gachsaran 5	5
Gachsaran 4	6
Gachsaran 3	7
Gachsaran 2	8
Gachsaran 1	9
Asmary	10
Pabdeh	11
Gurpi	12
Ilam	13
Sarvak	14
Fars	15

4.2 Machine Learning Model Application

Cross-validation involves re-sampling input data from multiple subsets and assessing the model's predictive performance in the overall dataset using a secondary subsets not utilised in the training process (Otchere et al., 2021a; Tarafder et al., 2021). This method may be used to minimise prediction errors on unknown data. The holdout cross-validation approach was used to randomly partition the data into 85:15 training and testing groups, representing 418 data points for testing. This approach guarantees that separate data points not observed by the trained model are used to validate the model leading to the avoidance of overfitting and selection bias. Several models were used in this research to effectively know the model that predicts mud loss with lower errors. The same out-of-sample data was used throughout this research to ensure that similar data points were used in all models.

Hyperparameter tuning of the final model was performed to further improve the model's performance. The BO algorithm, which is classed as a sequential model-based optimisation procedure, was used to improve the model (Tarafder et al., 2022). BO is a flexible and efficient approach that uses the Bayes Theorem to provide a rational mechanism for directing a global optimisation issue's investigation to the extrema of objective functions. The algorithm works by iteratively training a Bayesian approximation probabilistic model of the objective function based on prior outcomes approximation.

14 *Data Science and Machine Learning Applications in Subsurface Engineering*

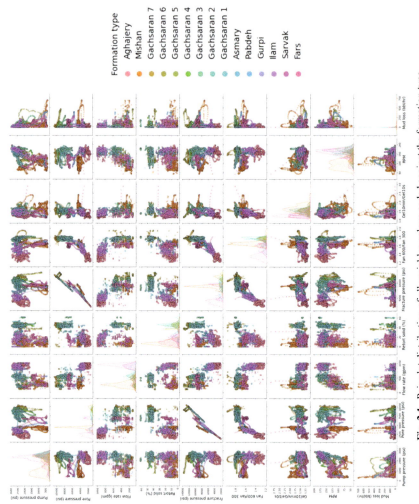

Fig. 2.1. Pair plot distribution of all variables colour-coded against the formation type.

The probabilistic model, a response surface that maps hyperparameters to the objective function probability score, is successfully evaluated with an acquisition function before selecting suitable samples to assess the actual objective function. Hyperparameter tuning is the act of updating the model's parameters to regulate the training procedure's performance and optimise the model's performance by achieving a reasonably lower cost function (Otchere et al., 2022c).

4.3 Explainable AI

In most cases, the application of feature selection techniques is unsupervised; hence, there is no right or wrong answer, making each technique different for each data. For this research, model agnostic methods will be used to analyse the importance of input features to separate its explanations from the machine learning model. The desirable characteristics sought after are model representation and explanation flexibility, which are not limited to a specific type and make sense in the context of the model being explained. For this purpose, the Permutation Feature Importance (PFI) and Shapley values will be used to explain the results generated by the models. The inputs to achieve this are the model, the feature vectors, the target, and the error metrics. After applying these techniques, features that do not explain the target will be removed from the input feature vector.

4.3.1 Permutation Feature Importance

Permutation feature importance (PFI) is a model agnostic approach for computing the significance of features (Otchere et al., 2022a). This approach works by permuting the relationship between a feature and the model output by assigning the feature a nearly random value. The basic working principle to achieve this is to compute the importance of a feature by measuring the increase in estimation error after fifty feature permutations. As a result, a feature is considered essential if omitting its value increases the model error. Otherwise, it is unimportant if the error remains constant (Breiman, 2001). The main advantages of applying PFI are its straightforward interpretation, highly compressed global insight into a model's behaviour, and not require retraining. However, one disadvantage is that it is unclear whether the training or test data should be used to measure PFI. The in-built function, permutation_importance, in Scikit-learn (Pedregosa et al., 2011) was used to run the PFI inspection technique. Given a regressor model, a function is created where the feature importance of all the input features is computed. An importance score criteria are computed whereby features achieving high values represent better predictive power.

4.3.2 Shapley Values

The Shapley value, introduced by Lloyd Shapley (Shapley, 1953) from Cooperative Game Theory, is a critical concept in measuring feature importance and the contribution of each feature to the models' performances. Shapley values work by allocating a unique distribution among the participants of the entire surplus value created by the coalition of all players in each cooperative game. The theory defines four properties that must be satisfied: efficiency, symmetry, dummy, and additivity. The players represent the input features in machine learning and interpretability, and the Shapley value determines how much each feature contributes to the output. Knowing how the features contribute to the model output helps in understanding the essential factors in generating the model's results. The Shapley value is not specific to any particular model type; hence it can be used regardless of the model type and architecture. Perhaps, unlike any other method of interpreting model results, Shapley's values are based on a solid conceptual framework. While making intuitive sense is essential for interpretability, other methods lack the same rigorous conceptual framework. Shapley values have one key advantage of being evenly distributed among an instance's feature values. It has been suggested that Shapley may be the only way to offer a complete explanation in cases when the law mandates it, such as the right to explanations. Shapley, like any other approach, has inherent drawbacks. The most crucial is that it is computationally expensive, implying that only an approximate solution can be calculated in a high percentage of real-world scenarios. It is also easy to misunderstand. After removing the feature from the model training, the Shapley value is not the difference between the expected and actual values. It is how much an input value contributes to the gap between the actual and average prediction.

The open-source Shap library is a robust tool for working with Shap values and other metrics. Shap establishes a link between efficient credit allocation and local explanations, making it model agnostic. It ties together optimum credit allocation and local explanations. Using game theory's traditional Shapley values and associated expansions have been the focus of numerous recent works. Recent academics have expanded on this notion by developing approaches for Shap values in deep learning models and gradient explainers. These approaches estimate Shap values for any model using specifically weighted local linear regression. It also provides various charts to help view the data and comprehend the model.

This study does not include the mathematical foundations of the models utilised because there is a vast collection of published material detailing them. An overview of the methodology workflow used in this study is demonstrated in Fig. 2.2.

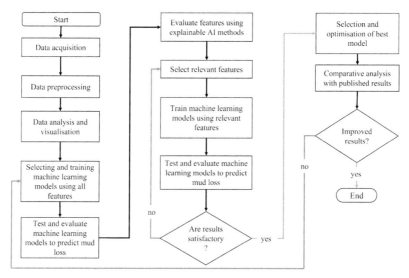

Fig. 2.2. Workflow summarising the methodology used in this study.

5. Results and Discussion

5.1 Evaluation of Model Performance

All of the models used in this study were evaluated using Akaike Information Criterion (AIC) to find the best fit for the data set. The assessment of the result shows that lower AIC values have a higher probability of fitting data compared to a larger AIC. The following factors were considered when comparing and selecting the best model. A difference in AIC findings across models is considered insignificant if it is less than 10, moderate between 10 and 50, significant between 50–100, and very strong if above 100. From the results, it can be asserted that the extra tree model ended up being the best for drilling fluid loss predictions. There was a difference of ~ 260 AIC when the extra tree model was compared to the random forest model representing the two models with the lowest AIC. Hence, the extra tree model showed a strong disparity and improved likelihood of fit compared to the other models. The tendency of the model to over- or underfit the data still exists, even though AIC can measure the degree to which the model converges to the data.

The models' performances were assessed based on the out-of-sample data results. Since the models were trained with the training data, any attempt to reproduce the values will likely result in high accuracy. On the other hand, a high training accuracy is not always desirable since it may overfit the data, collecting intrinsic noise and yielding a non-generalised model. Test accuracy based on the out-of-sample data is the proportion of correct predictions made by the models on data that has not yet been observed.

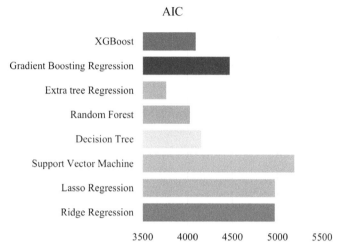

Fig. 2.3. AIC results comparing all the models on the test data.

Since the model aims to improve unseen data, it is particularly crucial that models have high test accuracy. In this work, correlation coefficient (R^2) values greater than 0.85 for both training and test accuracy scores were deemed a good trade-off over bias and variance. Table 2.4 shows that the Extra Tree model had th e highest R^2 on the test data for predicting mud loss. Figure 2.4 below depicts a comparison of all of the model's predictions to the actual data.

Any meaningful gain in accuracy, and hence a reduction in inaccuracy, achieved by machine learning models significantly influence decision-making. Each model's statistical and theoretical foundations resulted in widely variable performances. Figure 2.5 depicts the models' performances as a cross-validation evaluation to analyse their exactness, efficiency, and

Table 2.4. Train and test correlation coefficient score of all models used in this study.

Models	Train R^2 Score	Test R^2 Score
Extra Tree Regression	1	0.959819
Random Forest	0.984328	0.92394
XGBoost	0.99873	0.921476
Decision Tree	1	0.916176
Gradient Boosting Regression	0.868513	0.801631
Ridge Regression	0.372394	0.381738
Lasso Regression	0.367822	0.37421
Support Vector Machine	0.994892	−0.027344

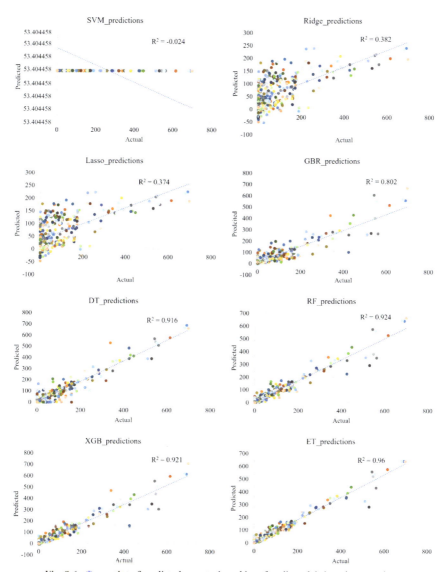

Fig. 2.4. Cross-plot of predicted vs actual mud loss for all models based on test data.

reliability. The Root Mean Squared Error (RMSE) findings for all the models indicate that the Extra Tree model evaluation is the most reliable compared to the actual mud loss values. The Mean Absolute Error (MAE) also suggests that the Extra Tree model is the most exact model for predicting mud loss. In selecting the best model, the ranking feature used in this study is as follows; MAE, RMSE, AIC, and R^2.

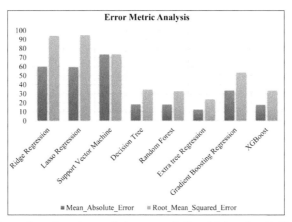

Fig. 2.5. Comparison of test data prediction errors of all models based on RMSE and MAE.

The performance of the extra tree model is attributed to the bias-variance concept used to build this model. The explicit randomisation of the cut-point and features combined with ensemble averaging can reduce variance more strongly than the weaker randomisation techniques employed by other algorithms. Aside from having similar advantages as Decision Trees based on consistency for universal generalisation and approximation, Extra Trees additionally give resilience regarding gross model errors because outliers affect its predictions slightly and locally. When the input features are significantly more than the random splits, Extra Trees potentially outperform Decision Trees relating to computational efficiency while resisting irrelevant input features.

5.2 Model Agnostic Results

Permutation importance based on the XGBoost model was computed to evaluate the importance of and analyse the effect of multicollinearity on the input variables. The results in Fig. 2.6 show that the pump and pore pressures highly correlate to the prediction of mud loss. The results indicate that permuting an input feature reduces the model's accuracy by, at most, 0.3, suggesting that some of the inputs are relevant. However, drilling time, hole size, meterage, WOB, and MFVIS do not offer any improvements to the model prediction. This result is compared with the feature importance output from the XGBoost model shown in Fig. 2.7. The discrepancy in the result indicates the importance of model agnostic that can give an inference into causality. This inference is necessary because correlation does not mean causation; hence, feature selection techniques based on correlation may not entirely improve model performance.

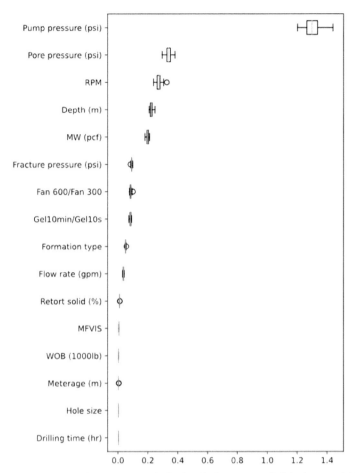

Fig. 2.6. Permutation importance plot indicating the importance of all the input variables.

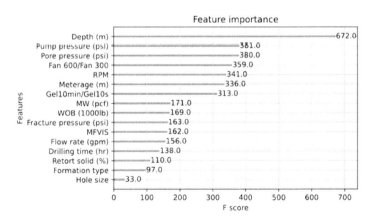

Fig. 2.7. XGboost Feature Importance of the input variables to the target.

5.3 Analysis of Features Using Model Agnostic Metrics

To maintain consistency in the final model prediction, a similar train/test split was employed, and the PFI results of the input features' performance on the entire data are presented in Fig. 2.8. The RF model yielded an initial testing R^2 of 0.92. The importance score is computed so that higher values represent a better predictive power. Generally, the importance score is calculated such that a higher value indicates a greater predictive power. A sizeable portion of the 0.92 accuracy score can be attributed to the importance of the values for the most relevant features. The results showed that some input variables were given a relatively low importance score. The results point to a small set of input variables that account for most of the predictability. Inferences can be drawn from this finding about the relative importance of the features. This result is in line with the high-test accuracy that was calculated.

The initial permutation importance calculated on the training data shows the model dependency on each feature during training. Nevertheless, it is essential to state that this analysis was done for both the training and testing dataset to help account for features that may help with the generalisation power of the model. Overfitting is more likely to occur when a feature is considered important during training but is dismissed as important during testing.

Sensitivity analysis was performed to select the relevant features for further prediction. The RF model was used to predict mud loss using the top

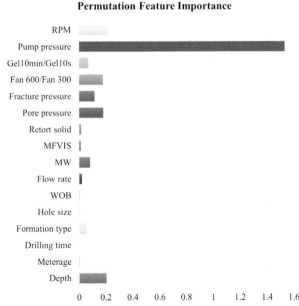

Fig. 2.8. Permutation feature importance of the input variables.

Table 2.5. Selection of input variables based on their permuting accuracy.

Features Selected	Features Dropped	Accuracy
16	None	0.9221
15	Drilling time	0.9238
14	Drilling time and Hole size	0.9226
13	Drilling time, Hole size, and Meterage	0.9348
12	Drilling time, Hole size, Meterage, and WOB	0.9325
11	Drilling time, Hole size, Meterage, WOB, and MFVIS	0.9304
10	Drilling time, Hole size, Meterage, WOB, MFVIS, and Retort solids	0.9339

15–10 features, and the result is illustrated in Table 2.5. The results indicate that the optimal number of features to predict mud loss is thirteen.

5.4 Analysis of Features Using Shapley Values Model Agnostic Metrics

The prediction of mud loss was performed using all the features to train the Extra Trees model. The same training and testing datasets used in the PFI analysis were used. Since machine learning models are supposed to be interpretable to help understand how a model made a particular prediction, model agnostic metrics are helpful. Shapley's values originated from the game theory, and it is necessary to clarify its application to supervised machine learning interpretability. The game represents the prediction assignment for a specific event in the dataset. The gain is the difference between the actual forecast for a specific instance and the average forecast for all instances. The instance's feature values collaborate to achieve the gain in the mud loss forecast.

Figure 2.9 visualises the shapely values as absolute values in a feature importance plot. The Shap feature importance works with a simple principle where the traits with high absolute values are the features of most importance. The absolute mean of the Shapley values for each attribute across the data is computed to evaluate the global importance. The pump pressure and flow rate feature variables were highly important, with pump pressure being the most important and changing the target on average by 56.35 bbl/hr on the x-axis. The plot also realised that the correlation of meterage, drilling time, and WOB to the target was low, with the inclusion of WOB disagreeing with PFI results. Based on this result, further investigation was necessary to ascertain which traits were irrelevant to the target. The importance of Shapley values is an alternative to the importance of permutation features. Both the importance measures have a significant difference. The decline in model performance determines the permutation attribute's relevance, whereas Shapley values are primarily based on the magnitude of feature provenances. The feature importance plot is

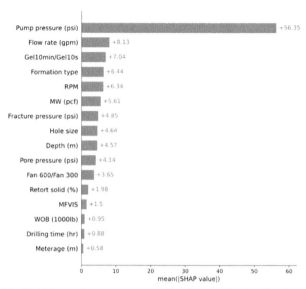

Fig. 2.9. SHAP feature importance measured as the mean absolute Shapley values.

informative but provides no other information outside the significance. The bee swarm technique, a more informative plot, is used for analysis.

Figures 2.10 and 2.11 visualise the Shapely values as absolute values in a bee swarm summary plot for both train and test data, respectively. The y-axis is determined by the feature, and the x-axis by the Shapley value. The feature importance with characteristic effects is combined into global explanations whereby a feature matrix of Shapley values is achieved for every instance. In the summary plot, the first indications of the positive and negative relationship between the value of a feature and the impact on the target is identified. The data points, made up of all the training data points, overlap in a scattered manner on the y-axis. This type of visualisation depicts the distribution of Shapley values per feature. The features are arranged in descending order of significance. From the plots, low feature values are represented by blue, while high feature values are denoted by red. In analysing the influence of pump pressure, it is observed that low values predict high mud-loss values, whereas high values predict low mud-loss values. The widespread of the pump pressure data points also indicates a global explanation that explains the entire model behaviour. This analysis confirms that this correlation on its own cannot be termed causation. This conclusion enforces the importance and need for model agnostic techniques to understand the influence the other features have in mud-loss prediction. From the bee swarm plot, it can be observed that about 12 features can globally explain how the predictions were made.

The results of the bee swarm plot confirmed that features that have high importance in both the train and test results and exhibit their importance in the

Enhancing Drilling Fluid Lost-circulation Prediction   25

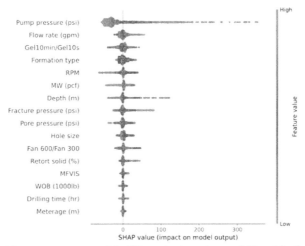

Fig. 2.10. A bee swarm summary plot of Shap value impact on model target for the train data.

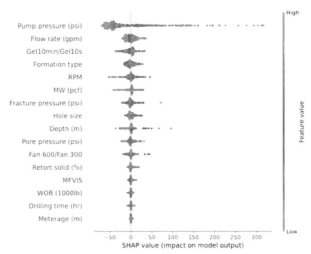

Fig. 2.11. A bee swarm summary plot of Shap value impact on model target for the test data.

global explanation of the target. From the demonstrated results, the following analysis was derived;

1. *Feature importance:* The features are ranked in descending order, and from the train and test data plot, the pump pressure is ranked first. The meterage feature has close to zero importance because it does not have any causal effect in predicting mud loss.
2. *Impact*: The horizontal location of the data points shows that the pump pressure has a negative correlation and a high prediction effect in general.

Shapley values are based on a solid theoretical foundation in game theory and compute feature values reasonably distributed to the target. This method also contributes to the unification of interpretable machine learning. The fast computation allows for the computation of multiple Shapley values required for the global model interpretations. Since the Shapley values represent the smallest element of the global interpretations, it makes global interpretations compatible with local explanations.

5.5 Evaluation of Top Features

Further analysis was performed using the top 5 performing models for the top 13 features based on PFI and Shapley values to ascertain if they will outperform the initial model predictions. PFI resulted in the removal of Drilling time, Hole size, and meterage. Meanwhile, Shapley values resulted in the removal of Drilling time, WOB, and Meterage. This result is shown in Table 2.6. All the top-performing models improved when the features based on PFI and Shapley values were used. However, it was noticeable that Shapley values selected the most relevant features, predominantly observed in the gargantuan improvement in model accuracy.

Table 2.6. Computed accuracy on test data using top 5 models.

Models	Accuracy Based on All Features	Accuracy Based on PFI Top 13 Features	Accuracy Based on Shapley Values Top 13 Features
Gradient Boosting	0.8016	0.8073	0.9999
Decision Tree	0.9162	0.9298	0.9990
Random Forest	0.9239	0.9290	0.997
Extra Tree	0.9598	0.9710	0.999
XGBoost	0.9215	0.9266	0.996

5.6 Model Optimisation

The BO algorithm was used to tune the model parameters to improve model performance. The results in Fig. 2.12 show the model performance of the extra tree model using the top 13 features identified by Shapley values and the BO–ET model. The results were compared to the results of Sabah et al. (2021). This comparison is necessary because the author used the same data as this study but employed different techniques and models in predicting fluid lost circulation. It is noteworthy to mention that although machine learning models depend on the data used, most of the data used in training the model in this study may not be entirely similar to the other study. However, it is expected that the majority of the data will be the same. As such, this comparison is mainly based on the use of the input features and how they

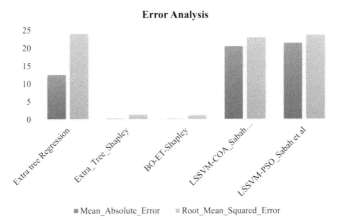

Fig. 2.12. Comparing the estimation errors of top-performing models and Sabah et al. (2021) models based on RMSE and MAE.

explain the target and not downgrade the results of the other researchers. From the error analysis performed, it was observed that the features selected based on the Shapley values significantly reduced the model's error. The initial extra tree model recorded an MAE of 12.6 bbls/hr, while the MAE from the Shapley selected features was 0.3 bbls/hr. This result represents about a 97% reduction in MAE.

Similarly, RMSE for the Extra Trees model using all 16 features was 24.0 bbls/hr. The RMSE from the Shapley selected features was recorded as 1.4 bbls/hr. Again, this represents about a 94% reduction in this error metric. Using hyperparameter tuning, BO–ET was able to reduce the extra tree's mean absolute error (MAE), and root mean squared error (RMSE) to 0.2 and 1.2 bbls/hr, respectively. The superiority of the Extra Trees model is mainly attributed to the bias-variance concept used to build this model, which makes it resilient to outliers. Based on all the evaluation criteria, it was determined that the Shapley selected features are highly relevant, offer a global generalisation of the target, and improve model efficiency.

5.7 Sensitivity Analysis

The kernel density estimate (KDE) of the estimated and actual mud-loss data is shown in Fig. 2.13. The estimated values are in blue, while the actual values are in green. According to the test data, the BO–ET is much closer to the actual data. This investigation indicates that the BO–ET can capture a broader range of values than the other models. This observation makes it suitable for use in evaluating drilling fluid lost circulation in other wells using the relevant drilling parameters. As a result, the sensitivity analysis verifies the assessment metrics outlined above.

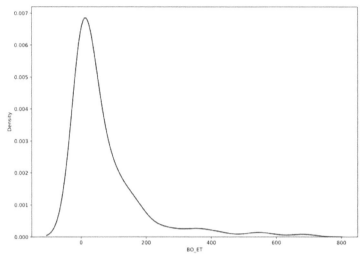

Fig. 2.13. Kernel density estimation for BO–ET model mud-loss prediction demonstrating the closeness of projected values to actual values.

The methodologies evaluated in this research demonstrated that explainable AI through the use of model agnostic metrics could give an insight into the input features. Since correlation does not mean causation, simple statistical techniques may not be able to indicate the causal effects of each input to the target. The findings show that gradually increasing the input variables above 13 causes a degradation in the model's performance. According to the Shapley values, drilling duration, meterage, and WOB are among the variables that have no direct influence on drilling fluid lost circulation. Hence, it should not be used to train the model. Model agnostic has provided trustworthiness and valuable solutions for predicting fluid loss using relevant causal features. The ability to adequately predict lost circulation prior to drilling gives vital information about fracture aperture and the optimal LCM particle-size distribution. As a result, LCM may be formulated to fill existing fractures or not to percolate deep into high permeable formations to minimise excessive drilling fluid loss. As such, predicting drilling fluid lost circulation can give valuable insight into the type of interventions required to be implemented.

6. Conclusions

This study demonstrated a methodology for identifying significant variables for drilling fluid lost-circulation estimations. The effectiveness of this approach has been assessed to surpass previously reported techniques based on the same dataset. The evaluation criteria discussed presented the probability of fitting held-out data of the models. The exactness, reliability, and dependability of the models were also assessed based on the R^2, MAE, and RMSE on the out-of-sample data. The models and how the target is predicted, based on the PFI and Shapley values, have made the input features explainable as the following traits are displayed;

1. *Fairness*: The results have ensured that the predictions are neutral and are not biased against any input variable, either implicitly or overtly. An easy-to-interpret model can explain why a particular prediction was made, making it more straightforward to determine which input variable had a more significant influence.
2. *Reliability*: The results help capture the differences in input variables that result in significant changes in the target and vice versa.
3. *Causality*: Causal relationships were identified, and the pump pressure feature for training and test data is ranked high amongst all input variables.
4. *Trust*: The results have yielded some level of trust in input features used for the final prediction. This analysis is based on how the employed techniques, PFI and Shapley values, explain their judgements on the input features to the target.

The application of Explainable AI enhanced model prediction accuracy by 97% and 94% in terms of MAE and RMSE, respectively. The comparison with other published results based on the same data resulted in a 98% and 95% reduction in MAE and RMSE, respectively. This analysis is proof that correlation does not mean causation. Without the PFI and Shapley values, the XGBoost feature importance would not have indicated features that will cause over- or underfitting and not capable of global interpretation of features.

Acknowledgement

The authors express their sincere appreciation to Universiti Teknologi Petronas, the Centre of Research in Enhanced Oil recovery, and the Centre for Subsurface Seismic Imaging for supporting this work.

Data Availability

The drilling data was obtained from Sabah et al. (2021).

References

Aalizad, S.A. and Rashidinejad, F. 2012. Prediction of penetration rate of rotary-percussive drilling using artificial neural networks: A case study/Prognozowanie postępu wiercenia przy użyciu wiertła udarowo-obrotowego przy wykorzystaniu sztucznych sieci neuronowych – studium przypadku. *Archives of Mining Sciences* 57: 715–728. https://doi.org/10.2478/v10267-012-0046-x.

Abbas, A.K., Al-haideri, N.A. and Bashikh, A.A. 2019a. Implementing artificial neural networks and support vector machines to predict lost circulation. *Egyptian Journal of Petroleum* 28: 339–347. https://doi.org/10.1016/J.EJPE.2019.06.006.

Abbas, A.K., Bashikh, A.A., Abbas, H. and Mohammed, H.Q. 2019b. Intelligent decisions to stop or mitigate lost circulation based on machine learning. *Energy* 183: 1104–1113. https://doi.org/10.1016/J.ENERGY.2019.07.020.

Abbas, A.K., Flori, R., Almubarak, H., Dawood, J., Abbas, H. and Alsaedi, A. 2019c. Intelligent prediction of stuck pipe remediation using machine learning algorithms. *In*: *SPE Annual Technical Conference and Exhibition*. SPE, Calgary. https://doi.org/10.2118/196229-MS.

Ahmadi, M.A. 2016. Toward reliable model for prediction Drilling Fluid Density at wellbore conditions: A LSSVM model. *Neurocomputing* 211: 143–149. https://doi.org/10.1016/j.neucom.2016.01.106.

Al-Baiyat, I. and Heinze, L. 2012. Implementing artificial neural networks and support vector machines in stuck pipe prediction. *In*: *Kuwait International Petroleum Conference and Exhibition*. SPE. https://doi.org/10.2118/163370-MS.

Alkinani, H.H., Al-Hameedi, A.T.T., Dunn-Norman, S., Flori, R.E., Alsaba, M.T., Amer, A.S. and Hilgedick, S.A. 2019a. Using data mining to stop or mitigate lost circulation. *J. Pet. Sci. Eng*. 173: 1097–1108. https://doi.org/10.1016/J.PETROL.2018.10.078.

Alkinani, H.H., Al-Hameedi, A.T.T., Dunn-Norman, S., Flori, R.E., Hilgedick, S.A., Al-Maliki, M.A., Alshawi, Y.Q., Alsaba, M.T. and Amer, A.S. 2019b. Examination of the relationship between rate of penetration and mud weight based on unconfined compressive strength of the rock. *J. King Saud. Univ. Sci.* 31: 966–972. https://doi.org/10.1016/j.jksus.2018.07.020.

Alkinani, H.H., Al-Hameedi, A.T.T. and Dunn-Norman, S. 2020. Artificial neural network models to predict lost circulation in natural and induced fractures. *SN Appl. Sci.* 2: 1980. https://doi.org/10.1007/s42452-020-03827-3.

Barbosa, L.F.F.M., Nascimento, A., Mathias, M.H. and de Carvalho, J.A. 2019. Machine learning methods applied to drilling rate of penetration prediction and optimization: A review. *J. Pet. Sci. Eng*. 183: 106332. https://doi.org/10.1016/j.petrol.2019.106332.

Brankovic, A., Matteucci, M., Restelli, M., Ferrarini, L., Piroddi, L., Spelta, A. and Zausa, F. 2021. Data-driven indicators for the detection and prediction of stuck-pipe events in oil & amp; gas drilling operations. *Upstream Oil and Gas Technology* 7: 100043. https://doi.org/10.1016/j.upstre.2021.100043.

Breiman, L. 2001. Random forests. *Mach. Learn.* 45: 5–32. https://doi.org/10.1023/A:1010933404324.

Chen, T. and Guestrin, C. 2016. XGBoost: A scalable tree boosting system. pp. 785–794. *In*: *Proceedings of the ACM SIGKDD International Conference on Knowledge Discovery and Data Mining*. ACM, New York, NY, USA. https://doi.org/10.1145/2939672.2939785.

Feng, Y. and Gray, K.E. 2017. Review of fundamental studies on lost circulation and wellbore strengthening. *J. Pet. Sci. Eng*. 152: 511–522. https://doi.org/10.1016/J.PETROL.2017.01.052.

Geurts, P., Ernst, D. and Wehenkel, L. 2006. Extremely randomized trees. *Mach. Learn.* 63: 3–42. https://doi.org/10.1007/s10994-006-6226-1.

Gordon, A.D., Breiman, L., Friedman, J.H., Olshen, R.A. and Stone, C.J. 1984. Classification and regression trees. *Biometrics* 40: 874. https://doi.org/10.2307/2530946.

Hoerl, A.E. and Kennard, R.W. 1970. Ridge regression: biased estimation for nonorthogonal problems. *Technometrics* 12: 55–67. https://doi.org/10.1080/00401706.1970.10488634.

Moazzeni, A., Nabaei, M. and Jegarluei, S.G. 2012. Decision making for reduction of nonproductive time through an integrated lost circulation prediction. *Petroleum Science and Technology* 30(20): 2097–2107. http://dx.doi.org/10.1080/10916466.2010.495961 30, 2097–2107. https://doi.org/10.1080/10916466.2010.495961.

Otchere, D.A., Arbi Ganat, T.O., Gholami, R. and Ridha, S. 2021a. Application of supervised machine learning paradigms in the prediction of petroleum reservoir properties: Comparative analysis of ANN and SVM models. *J. Pet. Sci. Eng.* 200: 108–182. https://doi.org/10.1016/j.petrol.2020.108182.

Otchere, D.A., Ganat, T.O.A., Gholami, R., Lawal, M., Arbi Ganat, T.O., Gholami, R. and Lawal, M. 2021b. A novel custom ensemble learning model for an improved reservoir permeability and water saturation prediction. *J. Nat. Gas Sci. Eng.* 91: 103962. https://doi.org/10.1016/j.jngse.2021.103962.

Otchere, D.A., Ganat, T.O.A., Ojero, J.O., Taki, M.Y. and Tackie-Otoo, B.N. 2021c. Application of gradient boosting regression model for the evaluation of feature selection techniques in improving reservoir characterisation predictions. *J. Pet. Sci. Eng.* 109244. https://doi.org/10.1016/J.PETROL.2021.109244.

Otchere, D.A., Abdalla Ayoub Mohammed, M., Ganat, T.O.A., Gholami, R. and Aljunid Merican, Z.M. 2022a. A novel empirical and deep ensemble super learning approach in predicting reservoir wettability via well logs. *Applied Sciences* 12: 2942. https://doi.org/10.3390/app12062942.

Otchere, D.A., Aboagye, M., Mohammed, M.A.A. and Boakye, T.B. 2022b. Enhancing drilling fluid lost-circulation prediction using model agnostic and supervised machine learning. *SSRN Electronic Journal*. https://doi.org/10.2139/ssrn.4085366.

Otchere, D.A., Ganat, T.O.A., Nta, V., Brantson, E.T. and Sharma, T. 2022c. Data analytics and Bayesian Optimised Extreme Gradient Boosting approach to estimate cut-offs from wireline logs for net reservoir and pay classification. *Appl. Soft Comput.* 120: 108680. https://doi.org/10.1016/j.asoc.2022.108680.

Pedregosa, F., Varoquaux, G., Gramfort, A., Michel, V., Thirion, B., Grisel, O., Blondel, M., Prettenhofer, P., Weiss, R., Dubourg, V., Vanderplas, J., Passos, A., Cournapeau, D., Brucher, M., Perrot, M. and Duchesnay, É. 2011. Scikit-learn: Machine learning in Python. *Journal of Machine Learning Research* 12: 2825–2830.

Pilehvari, A.A. and Nyshadham, V.R. 2002. Effect of material type and size distribution on performance of loss/seepage control material. *Paper presented at the International Symposium and Exhibition on Formation Damage Control, Lafayette, Louisiana.* pp. 863–875. *In*: *All Days*. SPE. https://doi.org/10.2118/73791-MS.

Sabah, M., Talebkeikhah, M., Agin, F., Talebkeikhah, F. and Hasheminasab, F. 2019. Application of decision tree, artificial neural networks, and adaptive neuro-fuzzy inference system on predicting lost circulation: A case study from Marun oil field. *J. Pet. Sci. Eng.* 177: 236–249. https://doi.org/10.1016/J.PETROL.2019.02.045.

Sabah, M., Mehrad, M., Ashrafi, S.B., Wood, D.A. and Fathi, S. 2021. Hybrid machine learning algorithms to enhance lost-circulation prediction and management in the Marun oil field. *J. Pet. Sci. Eng.* 198: 108125. https://doi.org/10.1016/j.petrol.2020.108125.

Shapley, L.S. 1953. A value for n-person games. pp. 307–318. *In*: *Contributions to the Theory of Games* (AM-28), Volume II. Princeton University Press. https://doi.org/10.1515/9781400881970-018.

Tarafder, S., Badruddin, N., Yahya, N. and Egambaram, A. 2021. EEG-based drowsiness detection from ocular indices using ensemble classification. pp. 21–24. In: *2021 IEEE 3rd Eurasia Conference on Biomedical Engineering, Healthcare and Sustainability (ECBIOS)*. IEEE. https://doi.org/10.1109/ECBIOS51820.2021.9510848.

Tarafder, S., Badruddin, N., Yahya, N. and Nasution, A.H. 2022. Drowsiness detection using ocular indices from EEG signal. *Sensors* 22: 4764. https://doi.org/10.3390/s22134764.

Tibshirani, R. 1996. Regression shrinkage and selection via the lasso. *Journal of the Royal Statistical Society: Series B (Methodological)* 58: 267–288. https://doi.org/10.1111/j.2517-6161.1996.tb02080.x.

Toreifi, H., Rostami, H. and Manshad, A.K. 2014. New method for prediction and solving the problem of drilling fluid loss using modular neural network and particle swarm optimization algorithm. *J. Pet. Explor. Prod. Technol.* 4: 371–379. https://doi.org/10.1007/s13202-014-0102-5.

Vapnik, V. and Lerner, A. 1963. Pattern recognition using generalised portrait method. *Automation and Remote Control* 24: 774–780.

Zarrouk, S.J. and McLean, K. 2019. Geothermal wells. *Geothermal Well Test Analysis* 39–61. https://doi.org/10.1016/B978-0-12-814946-1.00003-7.

CHAPTER 3

Application of a Novel Stacked Ensemble Model in Predicting Total Porosity and Free Fluid Index via Wireline and NMR Logs

Daniel Asante Otchere[1,2]

1. Introduction

With the growth in energy needs and advances in drilling and hydraulic fracturing techniques, the extraction of hydrocarbons from reservoirs innovatively has become a focus for exploration and exploitation in recent years (Li et al., 2022). Therefore, a complete understanding of bound and moveable fluids is critical for accurately evaluating net pay reservoirs and optimising completion, production, and enhanced oil recovery activities (Otchere et al., 2022b). In reservoir terms, pore fluids refer to the fluid distribution within the pore structure of the reservoir rock. There are two main types of pore volume fractions: bound fluid, the fluid that is physically trapped within the pore structure and cannot be easily displaced; and free fluid volume, the fluid that can easily move within the pore structure. Knowing the specific distinction between the proportion of bound and free fluid in a reservoir is essential for predicting the ease of fluid flow and production (Cai and Hu, 2019). For example, a reservoir with a higher proportion of bound fluid may require more aggressive enhanced recovery methods to increase

---

[1] Centre of Research for Subsurface Seismic Imaging, Universiti Teknologi PETRONAS, 32610, Seri Iskandar, Perak Daril Ridzuan, Malaysia.
[2] Institute for Computational and Data Sciences, Pennsylvania State University, University Park, PA, USA.
Email: ascotjnr@yahoo.com

production. Understanding pore fluid type and volume fractions is crucial for effective reservoir management and optimisation (Cai and Hu, 2019; Otchere et al., 2021c).

Numerous approaches have been formulated to ascertain the fluid distribution in a reservoir. The most appropriate and accurate method to measure pore types in reservoirs depends on the specific characteristics of the reservoir being studied. One standard method is the mercury injection capillary pressure (MICP) analysis, which measures the amount of mercury that can be injected into the pore structure at a given pressure (Peng et al., 2017). The amount of mercury that can be injected into the pore structure is directly related to the proportion of free fluid in the reservoir. MICP is a widely used method for measuring fluid distribution, and it is considered one of the most accurate and reliable methods for measuring free fluid volume (Dugan, 2015; Mitchell et al., 2008). Another method is nuclear magnetic resonance (NMR), which can provide a detailed description of the pore structure and fluid distribution within the reservoir (Xu et al., 2015). NMR is particularly useful for complex pore structures and can provide information on the distribution of multiple fluid phases. Additionally, NMR is a non-destructive method that uses magnetic fields and radio waves to measure the magnetic susceptibility of the pore fluids (Branco and Gil, 2017; Otchere et al., 2022a). The magnetic susceptibility of the pore fluids is used to distinguish between the bound and free fluid in the reservoir, making it a valuable tool for measuring pore fluid types and volume fractions on a large scale (Heaton et al., 2000).

Both methods, however, have their inherent set of challenges. A fundamental challenge in measuring or determining pore type is the complexity of the pore structure. Reservoirs can have a variety of pore sizes and shapes, and different methods may be more effective for different types of pore structures. One challenge associated with using MICP is that it requires a core sample from the reservoir, which can be costly and time-consuming. Additionally, the availability of core samples may be limited, making it difficult to obtain enough samples to provide a representative analysis of the reservoir (Wu et al., 2021). Also, MICP is a destructive method, where the core sample is destroyed after analysis, which could make it difficult to repeat the analysis for the same sample if needed. Another challenge is the sensitivity of MICP to the saturation of the fluid, requiring the fluid to be in a state of equilibrium, which may not be the case in some reservoirs. This challenge can affect the accuracy of the results obtained from MICP analysis (Wu et al., 2020). On the other hand, NMR is sensitive to the reservoir's porosity and cannot provide quantitative measurements of pore types. Furthermore, NMR is responsive to the existence of clay minerals, which can impact the precision of the outcomes (Elsayed et al., 2020). Table 3.1 summarises the unique advantages and limitations of these two techniques.

Table 3.1. Summary of MICP and NMR techniques identifying key advantages, limitations, and measurement principles.

Technique	Measurement Principle	Developed by	Advantages	Limitations	Observations
Mercury Injection Capillary Pressure (MICP)	Injects mercury into the pore structure at a given pressure, and the amount of mercury that can be injected into the pore structure is directly related to the proportion of free fluid in the reservoir.	Purcell (1949)	Accurate measurement of free fluid volume (FFV)	Requires a core sample	MICP is considered one of the most accurate and reliable methods for measuring FFV, but it is a destructive method.
Nuclear Magnetic Resonance (NMR)	Uses magnetic fields and radio waves to obtain detailed images of the fluid distribution within the reservoir.	(Bloch, 1946)	Provides detailed images of the pore structure and fluid distribution within the reservoir.	Not able to provide quantitative measurements of pore types and sensitive to the presence of clay minerals.	NMR imaging is particularly useful for complex pore structures and can provide information on the distribution of multiple fluid phases.

Overall, understanding pore volume fraction is crucial for effective reservoir management and optimisation. However, accurately measuring and determining porosity and its associated producible volume can be challenging due to the complexity of the pore structure and the limited availability of core samples.

The measurement of pore volume fraction is complicated and challenging. Many researchers in this discipline are always experimenting with new ideas and methodologies. Many researchers have been drawn to estimate pore volume fractions and total porosity using NMR data over the last two decades. This preference for NMR is owing to its non-invasive nature (Otchere et al., 2022a). Artificial intelligence (AI) can potentially revolutionise how pore volumes are predicted in reservoirs. One potential application is using AI to predict pore volumes measured by NMR, using wireline logs as input.

Conventional wireline logs are ubiquitous in all wells drilled and provide essential information on the formation properties of the reservoir. These logs exhibit an intricate and nonlinear correlation with the distribution of pore volume fractions. On the other hand, NMR measurements provide information on the fluid distribution within the pore structure. Nonetheless, this approach incurs significant expenses and may not appear practical to execute in every well during periods of low oil prices (Otchere et al., 2022a). By combining these two types of data, AI can be used to predict porosity and volume fractions with a high degree of accuracy. An ideal way to accomplish this is through the use of machine learning and data analytical techniques, which can be trained on a large dataset of wireline logs and NMR porosity and volume fraction measurements. These techniques exhibit resilience to noise and demonstrate a high efficacy in identifying intricate or nonlinear patterns or features indicative of specific input data and pore volume fraction (Otchere et al., 2021a). Once trained, the AI model can then be used to predict total porosity and pore volume fractions for new reservoirs based on the wireline logs.

Utilising wireline logs to assist in estimating this indispensable property is comparatively more economical and time-saving and provides real-time information than running NMR logs in all wells. As such, considering current oil prices, the demand for an inexpensive and reliable field-scale technique for quantitative porosity and pore volume fraction measurement has been of interest. Hence, this study endeavours to develop an AI-based model that can accurately predict the porosity and producible pore volume fraction in a carbonate gas reservoir using wireline logs as input and NMR measured total porosity and FFV as the target output. This study aims to provide a more efficient and accurate method for reservoir characterisation and management.

2. Nuclear Magnetic Resonance

2.1 Concept and Application

Atomic nuclei with an odd number of protons and neutrons exhibit NMR when subjected to magnetic fields at a specific resonance frequency (Coates et al., 1999). The proton in hydrogen has a strong magnetic moment due to its abundance in both water and oil, which is the outcome of the nucleus' angular momentum. At that point, the proton acts like a spinning bar magnet. In an external magnetic field, it undergoes polarisation, which manifests as a quantifiable quantum feature known as the induced frequency (Larmor frequency) (Amani et al., 2017). Figure 3.1 shows a simplified representation of the reaction of a nucleus to an external magnetic field (Otchere et al., 2022a).

Combinable Magnetic Resonance (CMR) tool is a recently developed technique that is used for porosity partitioning to determine the bound fluid volume (BFV) and FFV in reservoirs (Gubelin and Boyd, 1997). CMR can be used in conjunction with NMR to provide a more accurate and complete picture of the fluid distribution within the reservoir. CMR measurements are particularly useful in reservoirs with high water saturation, where other techniques may not be effective. Additionally, CMR can provide information on both bound and FFVs, which is not possible with other techniques.

Application of a Novel Stacked Ensemble Model in Predicting Total Porosity 37

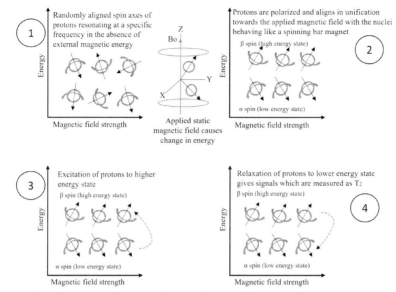

Fig. 3.1. The excitation and relaxation of polarised hydrogen atomic nuclei in response to an external magnetic field (Otchere et al., 2022a).

The concept of using NMR to predict pore fluid types is to use the magnetic properties of certain nuclei to provide information about a reservoir's fluid composition and rock properties. This prediction is achieved by measuring the proton density and relaxation times of the fluids in the reservoir, which are unique to different fluid types, such as water, oil, and gas. When CMR is used for porosity partitioning, the total porosity is calculated by measuring the magnetic susceptibility of the pore fluids and using the relationship between magnetic susceptibility and fluid volume fraction. This method is based on the assumption that the magnetic susceptibility of the bound fluid is different from the free fluid, which is a key concept in CMR. The following equations are used to calculate total porosity, BFV, and FFV in reservoirs:

$$(\phi T) = \phi B + \phi F \qquad (3.1)$$

where the total porosity is represented as ϕT, which is a dimensionless value between 0 and 1, representing the fraction of the rock sample that is made up of pores, ϕB is the bound fluid volume fraction, and ϕF is the FFV fraction.

$$BFV = \phi B \times V_p \qquad (3.2)$$

where V_p is the pore volume.

$$FFV = \phi F \times V_p \qquad (3.3)$$

V_p is typically calculated by multiplying the bulk volume of the rock sample (Vb) by the total porosity (ϕT).

$$V_p = \phi T \times V_b \tag{3.4}$$

where V_b is the bulk volume of the rock sample, which can be measured by using various techniques such as helium pycnometry, mercury intrusion porosimetry, or gas adsorption. It is worth mentioning that the pore volume can also be calculated by measuring the rock sample's bulk volume and subtracting the solid grains' volume from it. This method is known as the grain volume method and is another way to calculate the pore volume. These equations are based on the assumption that the magnetic susceptibility of the bound fluid is different from the free fluid. This means that the bound fluid can be distinguished from the free fluid by measuring the magnetic susceptibility of the pore fluids.

2.2 Works Related to the Use of Machine Learning in NMR for Reservoir Characterisation

Machine learning has been extensively employed to complement conventional methods and, in some instances, provide certain advantages. The deployment of machine learning techniques and AI algorithms to overcome intricate issues has gained traction in numerous sectors, including the petroleum industry (Otchere et al., 2021a). The application of machine learning in reservoir characterisation has seen massive growth underlining its importance in this field. Machine learning models have become quite useful in situations where not much data is collected. These models have been used to predict various logs, which, due to time and cost, were not run in wells. These models have illustrated high accuracy compared to actual logs, saving time and money while providing a real-time estimate of reservoir properties during drilling. A review of articles utilising machine learning techniques and NMR data to provide accurate alternatives is presented below.

Rezaee (2022) expounds upon a study that employed an extensive repository of CMR to engender outputs from an NMR logging tool for clastic rocks. The study intended to evaluate the effectiveness of diverse machine learning (ML) methodologies employing customary well-logs as input data. The study generated outputs for CMR data, including irreducible pore fluid (BVI), clay bound water (CBW), producible fluid (FFI), irreducible water saturation (Swirr), logarithmic mean of T_2 relaxation time (T_2LM), and permeability. To this end, well logs from 14 Western Australian wells were gathered, of which 80% were used for model training and validation and the remaining 20% for testing purposes. Upon comparing the outcomes, it was revealed that the Adaptive Boosting (AdaBoost) model demonstrated the most notable results, with R^2 values exceeding 0.9 for the blind set. These findings indicate that the ML methodology can be employed for the generation of NMR logs with a high degree of precision.

Masroor et al. (2022) describe an ML procedure for estimating NMR permeability from conventional well logs. The methodology employs three supervised (ML) algorithms: group method of data handling, Random Forest (RF), and one-dimensional convolutional neural network (CNN). Additionally, a modified version of the two-dimensional CNN (Residual 2D-CNN) is devised, receiving artificial 2D feature maps derived from conventional logs. The ML and deep learning (DL) models' hyperparameters are optimised using genetic algorithms to enhance their efficacy. The results indicate that nonlinear machine and DL techniques help estimate NMR permeability, with the Res 2D-CNN model providing the most efficient results (accuracy of 0.97). The study underscores the significance of employing produced feature maps to train the Res 2D-CNN model and the favourable impact of integrating residual and deeper bottleneck architectures in enhancing prediction accuracy and minimising training duration.

Tamoto et al. (2023) also describe a study aimed at developing ML models for the estimation of NMR porosity from well logs. NMR logs are considered valuable for porosity quantification but are limited by high costs and adverse subsurface acquisition conditions. The adjusted RMSE and R^2 and metrics compared and evaluated four supervised ML models. The best results were achieved with the CatBoost regressor, producing an RMSE of less than 0.01 and an adjusted R^2 of 0.87. The ML models substantially improved total porosity prediction compared to traditional empirical computations. The dissimilarities between actual NMR logs and the ML predictions were mostly below 5%. Additionally, a porosity model based on well logs was developed using tree boosting, and the impacts of input variables on model estimates were investigated. The study also scrutinised the behaviour of the model's linear and nonlinear characteristics to comprehend their relationships with the dataset.

Li et al. (2020) present a study on the use of two neural-network-based ML models, long short-term memory (LSTM) network and a variational autoencoder with a convolutional layer (VAEc) network, for the generation of synthetic NMR T_2 distributions from formation mineral and fluid saturation logs. The models are trained and tested on a limited sample of data from the Bakken shale formation. The results show that both models perform well, synthesising fluid-filled pore size distributions even when the input data is corrupted by noise. The proposed method has the potential to improve reservoir characterisation under data constraints. NMR logs provide valuable information on fluid mobility, fluid-filled pore size distribution, fluid mobility, porosity, and permeability in the near-wellbore reservoir volume. However, their acquisition is limited due to financial and operational challenges. Hence, the use of ML can make NMR logs available in all wells.

Gu et al. (2021) present a ML approach to estimate fluid saturation in reservoirs by evaluating fluid components in the T_2-D spectrum. The method integrates a GABP neural network and Gaussian Mixture Model (GMM) to determine a new saturation formula based on the physical significance between the T_2-D spectrum and fluid saturation. The approach is evaluated through numerical simulations, and results demonstrate improved accuracy and stability compared to traditional BP neural networks. The GMM determines different Gaussian probability distribution functions to characterise different fluids in the T_2-D spectrum, leading to more accurate saturation estimates. The proposed method has the ability to quantitatively evaluate fluid components in the T_2-D spectrum.

These studies demonstrate the potential of using ML algorithms to analyse wireline log and NMR data and predict various reservoir properties such as pore-scale properties, permeability, fluid saturations, and fluid types. These methods have been shown to be accurate and have the potential to improve the efficiency and accuracy of reservoir characterisation.

3. Methodology

3.1 Data Collection and Description

Conventional wireline logs from a vertical exploration gas well were used in this study. The CMR tool was logged in the well. The objective of the CMR logging was to evaluate the formation's total porosity, bound and free fluid, and permeability. CMR was logged over a carbonate section capped by shale. Over the zone of interest, we do not observe borehole effects on the CMR signal due to a relatively smooth borehole combined with moderate mud salinity (approximately 58 ppk equivalent NaCl using chart GEN-9 with Rmf = 0.0694 ohmm at 21 deg C). No anomalies were observed on the raw echoes, T2 distribution, total porosity, and bin porosities. Since the well is drilled with water-based mud, the T_2 distribution reflects the pore size distribution. The data used in this study underwent quality control and reprocessing. The parameters selection for the CMR reprocessing used in this study is captured in Table 3.2. Figure 3.2 shows the incremental increase in porosity and multiple pore size distribution with favourable connectivity between pores, identified through the multi-exponential decay time analysis of NMR T_2 distribution in 100% brine saturated samples. The partitioning of the porosity was done as follows:

1. *Small Pore Porosity* (T_2 min to 3 ms) – this can be considered as water-filled porosity associated with clay (water resistivity of Rwb in the Dual Water Saturation model).
2. *Capillary Bound Fluid* (3 ms to T_2 cutoff) – this can be considered part of effective porosity. In clastic rocks, this would be expected to be water

Table 3.2. Processing parameters selection for CMR reprocessing.

Processing Parameters	
Porosity Algorithm	
Starting Echo	2nd
Number of averaging levels	3
T_2 minimum	0.5 msec
T_2 maximum	6000 msec
Number of Spectral Components	30
T_1/T_2 Ratio Minimum	1
T_1/T_2 Ratio Maximum	3
Polarization Correction Threshold	0.015 v/v
Producibility Parameters	
Free Fluid Cut-off	0.02 v/v
Porosity Parameters	
T_2 Cut-off	100 msec (default for Carbonate)
Taper Cut-off Start	8 msec
T_1/T_2 Ratio	2
Bin Porosity T_2 Cutoffs (ms)	1, 3, 10, 33, 100, 300, 1000, 3000
Small Pore Porosity T_2 Cutoffs (ms)	1 to 3
Capillary Bound Fluid T_2 Cutoffs (ms)	3 to 100

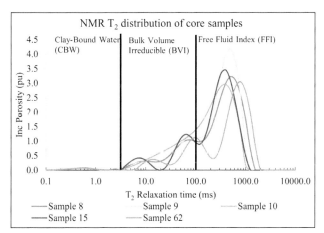

Fig. 3.2. NMR T_2 multimodal pore size distribution of different core samples at 100% water saturation (Otchere et al., 2022a).

filled (with water resistivity equal to Rw in the Dual Water Saturation model); however, in carbonate rocks, this could be filled with water, hydrocarbon or both.

3.2 Data Analysis and Feature Engineering

The relationship between output and input features was established by computing a covariance heatmap and pair plots through Jupyter notebook, a Python-based computing platform. Tables 3.3 and 3.4 present the statistical summary of the data, aiding in identifying the data distribution and expected trends.

These logs were chosen due to their established relationship to porosity using the studies by Otchere et al. (2021b). In the next stage, several ensemble and DL models were used to predict the porosity and FFV using the selected wireline logs.

Table 3.3. Descriptive statistics of the input data used in this study.

	S-sonic	Gamma	SP	R(S) HLLS	Rt	Rxo	Neutron	RHOB	SXGAS
mean	162.64	41.78	−388.62	102.37	335.15	4.33	0.14	2.33	0.51
std	39.73	19.45	27.70	259.54	1628.80	3.84	0.06	0.20	0.35
min	99.98	24.57	−431.31	1.45	1.32	0.86	0.02	1.76	0.00
25%	138.79	31.47	−411.00	12.17	12.44	2.09	0.10	2.17	0.22
50%	149.13	34.83	−397.50	35.94	46.82	3.21	0.13	2.37	0.56
75%	172.14	41.96	−369.56	111.24	180.35	5.19	0.16	2.47	0.82
max	309.04	114.09	−314.13	4986.61	28651.10	51.44	0.37	2.66	1.00

Table 3.4. Descriptive statistics of the targets used in this study.

	TCMR	CMFF
mean	0.11	0.07
std	0.03	0.04
min	0.01	0.00
25%	0.09	0.04
50%	0.11	0.07
75%	0.13	0.09
max	0.31	0.16

3.3 Machine Learning Model Application

It is widely accepted that DL algorithms perform better than their traditional, more simplistic equivalents in the field of ML. The ability of DL models to manage substantial amounts of data has resulted in their widespread popularity for tasks involving predictive modelling and pattern recognition (Otchere et al., 2021a). Despite all of the advancements that have been made thanks to DL in the field of AI, it still has significant flaws. Hyperparameter tuning, slow convergence on short datasets, infinite structures, and complexities are among the most often encountered problems. While traditional ML models can mitigate these issues on smaller datasets, they are limited in their ability to produce higher model performance overall. It is difficult to know exactly which model to use because of the hazy definition of small datasets. Based on these assertions, a new custom stacked ensemble model is developed in this research. This custom ensemble model integrates multiple varied base learners in its architecture to enhance model effectiveness. However, compared to the custom stacked ensemble model, the performances of the individual models are low.

3.3.1 Building Deep Learning Models

A CNN model is a type of DL model that is particularly well-suited for image and signal processing tasks. However, its performance was evaluated on a regression task in this study. It comprises multiple layers, including convolutional, pooling, and fully connected layers. To build a CNN model for this study, the following steps were taken:

1. The input data, wireline logs with 10 dimensions, were reshaped to match the expected input shape for a CNN model.
2. The model was composed of a series of convolutional layers designed to extract features from the input data. These layers use filters to scan over the input data and extract features at different scales.
3. The output of the convolutional layers was then passed through one or more pooling layers, which minimised the dimensionality of the feature maps and offered some form of translation invariance.
4. The final output of the pooling layers was then passed through one or more fully connected layers to make the final predictions.
5. The model was then trained using the training dataset and evaluated using the holdout dataset.
6. The model was fine-tuned by adjusting the number of layers, the number of filters in each layer and the size of the filters to achieve the desired level of performance.

7. The model was trained using the Adam optimiser with MAE as the loss function.
8. The model was trained using early stopping to prevent overfitting.

Long Short-Term Memory (LSTM) and Gated Recurrent Unit (GRU) models are types of Recurrent Neural Network (RNN) models that are particularly well-suited for sequential data such as time series, speech, and text. A GRU model is similar to an LSTM model but has fewer parameters, which can be useful when working with limited data or computational resources. The key feature of GRU and LSTM is the ability to remember information over a more extended period of time by using gates to control the flow of information in and out of the cell. This model type can be useful when working with sequential data where the order of the data is essential. However, this model type can be computationally intensive and requires a large amount of data for training. To build the GRU and LSTM model for this study, the following steps were taken:

1. The input data was transformed into a suitable format for GRU/LSTM, usually in the form of a 3D array where the first dimension represents the number of samples, the second dimension represents the time steps, and the third dimension represents the number of features.
2. The model was composed of one or more GRU/LSTM layers, which are responsible for processing the input data and extracting features.
3. The output of the GRU/LSTM layers was then passed through one or more fully connected layers to make the final predictions.
4. The model was then trained using the training dataset and evaluated using the holdout dataset.
5. The model was fine-tuned by adjusting the number of layers, the number of neurons in each layer, and the dropout rate to achieve the desired level of performance.
6. The model was trained with regularisation techniques such as dropout and early stopping to prevent overfitting.

3.3.2 Building a Hybrid Stacked Ensemble Model

A hybrid ensemble model is a combination of multiple models working together to improve the overall performance of the model. A hybrid ensemble model of Extra Trees, Random Forest, and XGBoost was built using stacking in this case. Stacking is a technique where the predictions of the base models (Extra Trees, Random Forest, and XGBoost) are used as input to train a meta-model, which makes the final prediction. The base models were trained on the same dataset, and their predictions were combined into a single array,

Table 3.5. Model architecture stacking ensemble models as base learners for the Custom Ensemble. Adapted from Otchere et al. (2022a).

Algorithm 1: Stacking Base Learners
1: **def** ensemble_predict(models, X_train, y_train, X_test):
2: # create an empty array to store predictions
3: predictions = list()
4:
5: # loop through the models
6: for model in models:
7: model.fit(X_train, y_train)
8: y_pred = model.predict(X_test)
9: predictions.append(y_pred)
10:
11: # return the predictions
12: return predictions
13:
14: **def** train_stacking_ensemble(X_train, y_train, X_test, y_test):
15: models = get_models()
16: ensemble_predictions = ensemble_predict(models, X_train, y_train, X_test)
17: meta_model = LinearRegression()
18: meta_model.fit(ensemble_predictions, y_test)
19:
20: # predict on the test set
21: y_pred = meta_model.predict(ensemble_predictions)
22:
23: # evaluate performance
24: r^2 = r2_score(y_test, y_pred)
25: mae = mean_absolute_error(y_test, y_pred)
26: rmse = sqrt(mean_squared_error(y_test, y_pred))
27: fit_likelihood = OLS(y_test, add_constant(y_pred)).fit()
28: print("R^2: {:.4f}".format(r^2))
29: print("MAE: {:.4f}".format(mae))
30: print("RMSE: {:.4f}".format(rmse))
31: print("AIC: {:.4f}".format(fit_likelihood.aic))

which was used as input to train a meta-model using linear regression. The meta-model was trained on the combined predictions and the target variables of the training dataset, and it was then used to make predictions on the test set. This approach can improve the model's performance by combining the strengths of different models and reducing the prediction variance (Tarafder et al., 2022). Table 3.5 summarises the function used to develop the stacked ensemble using the Extra Trees, Random Forest, and XGBoost models, whiles Fig. 3.3 summarises the entire workflow.

There are several different combinations of models and different parameters for each model, as well as different meta-models to find the best hybrid ensemble for a specific task, dataset, and desired performance.

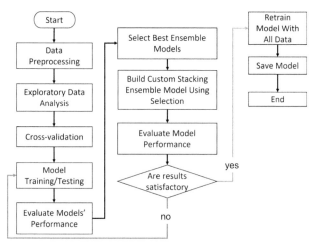

Fig. 3.3. Simplified workflow of the methodology used in this study.

3.4 Criteria for Model Evaluation

The discrepancy between estimated and actual data shows the models' prediction errors when assessing their performance. The performance of the models in this study was analysed via these evaluation criteria;

1. *Mean Absolute Error (MAE)*: This assessment criterion, also known as L1 loss, is vulnerable to relative errors and resistant to global scaling of the expected outcome. MAE is mathematically written as:

$$MAE = \frac{\sum_{i=1}^{n}|y_i - \hat{y}_i|}{n} \quad (3.5)$$

2. *Root Mean Squared Error (RMSE)*: This assessment criterion is the standard deviation of models' prediction errors and indicates how close predicted data is to actual data. This is written as:

$$RMSE = \sqrt{\frac{\sum_{i=1}^{n}(y_i - \hat{y}_i)^2}{n}} \quad (3.6)$$

3. *Akaike Information Criterion (AIC)*: This assessment metric is based on a frequentist probability approach, which scores a model according to its maximum probability estimation. This technique is used to determine the precision and excellence of models, indicating a more robust match between the model and new data. AIC is statistically written as:

$$AIC = 2K - 2(log - likelihood) \quad (3.7)$$

4. *Coefficient of determination (R^2)*: R^2 denotes the closeness of the dependent value to the best-fit regression line. R^2 is mathematically expressed as:

$$R^2 = 1 - \frac{\sum(y_i - \hat{y}_i)^2}{\sum(y_i - \bar{y}_i)^2} \qquad (3.8)$$

Besides R^2, low errors are indicative of good model performance.

4. Results and Discussion

4.1 Evaluation of Models' Performances

Given the expense associated with operating an NMR log and the efficacy of machine learning models in reservoir characterisation, a variety of models were employed to estimate the total porosity and free fluid volume of a carbonate reservoir from wireline logs. The findings in Fig. 3.4 indicate that the hybrid ensemble exhibited an improved accuracy of holdout data in the prediction of total porosity and moveable fluid compared to the other models. Although some models, Extra Trees and XGBoost, achieved the highest training accuracy but comparatively lower test accuracy compared to the hybrid model, the models likely overfitted to the training data. Overfitting occurs when a model is too complex and can fit the training data very well but cannot generalise well to the test data. This issue is usually caused by having too many parameters or too much capacity in the model, which allows it to fit the noise in the training data. In this case, the Extra Trees model has a high training accuracy of 1, which is an indication that it is able to fit the training data very well. However, the test accuracy of 0.82 is lower than

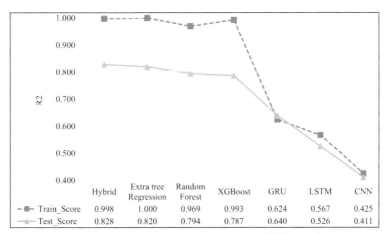

Fig. 3.4. Comparison of train/test correlation coefficient results.

the hybrid model, which indicates that the Extra Trees model is not able to generalise well to new data. On the other hand, the hybrid model combines the predictions from multiple models, which reduces the variance and increases the robustness of the final prediction. Therefore, it is less likely to overfit the training data and more likely to generalise well to new data. Based on this alone, it is vital to note that the hybrid model performed more robustly than the other models. Additional analysis was conducted to assess the prediction errors of all the models.

Error measurements were used to assess the consistency and precision of the models and are presented in Fig. 3.5. The MAE and results indicate the hybrid model's superiority over other models, indicating its consistency and precision in multioutput prediction. The Extra Trees Regression model's performance was comparative, with an RMSE of 0.0143 and an MAE of 0.0102. Its AIC score is also higher than the hybrid ensemble model indicating that the hybrid model has a higher probability of fitting new data. The Random Forest and XGBoost models performed relatively well, with RMSE of 0.0.0152 and 0.0.0155, respectively. Their MAE was higher than the hybrid ensemble model.

On the other hand, the LSTM and GRU models performed poorly compared to the other base models, with RMSE of 0.0231 and 0.0202 and MAE of 0.0176 and 0.0151, respectively. The AIC scores are also higher than the other models, indicating that they are less likely to generalise well to new data. Comparing the two RNN models indicate that the GRU model recorded lower errors than the LSTM model. The worst performing model, the CNN model, achieved an MAE of 0.0273 and RMSE of 0.034, indicating that it is not able to fit the data well. The relatively high AIC score also indicates that it is not the best model for this task.

The results suggest that the hybrid ensemble model is the best performing model for this study, as it has the highest train and test scores, the best AIC score, and the lowest MAE and RMSE. The Extra Trees Regression, Random Forest, and XGBoost models performed well, but some overfitting issues were observed. The ensemble models should be further investigated and optimised to improve their performance. Moreover, it would be interesting to explore other ensemble methods, such as bagging and boosting, to improve the model performance. Furthermore, a larger dataset and more computational resources may be necessary to enhance the performance of the LSTM and GRU models (Fig. 3.5).

Figures 3.6 and 3.7 below illustrate the kernel density estimation (KDE) of the Hybrid, GRU, and ET predicted and actual total porosity and FFV. From the observation, the hybrid ensemble predicted outputs are significantly closer to the total porosity and free fluid volume actual data. This result suggests that the hybrid model is better equipped to capture a wide range of values

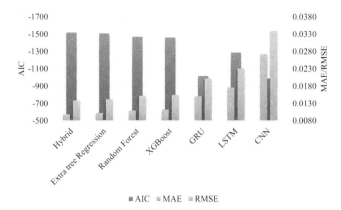

Fig. 3.5. Comparison of the models' estimation errors based on RMSE, MAE, and AIC on the holdout data.

compared to the other models and is suitable for deployment in predicting total porosity and FFV. A joint plot in Fig. 3.8 was used for visualisation to further evaluate the hybrid model's performance. The shaded area on the plot shows a 95% confidence interval, which indicates high confidence in the predicted total porosity and FFV values when compared to the actual. The low confidence area is mainly in the prediction of low values as a result of the low porosity values in the training data. The low values are due to the data restriction in the reservoir interval, which will mostly have high porosity values. Thus, the previously mentioned evaluation metrics results are verified by conducting this sensitivity analysis.

The excellent performance of the hybrid model is mainly attributed to its learning theory, model architecture, and mathematical formulation. From a learning theory perspective, the hybrid model combines the strengths of multiple base models, such as Extra Trees Regression, Random Forest, and XGBoost, all ensemble methods known to enhance a single model estimation performance by reducing overfitting and increasing generalisation. This is achieved by combining predictions from multiple models, which reduces the variance and increases the robustness of the final prediction. From a model architecture perspective, the hybrid model uses a stacking ensemble technique to combine the base models. This technique involves training multiple base models on the input data and then using their predictions as input to a higher-level meta-model that makes the final predictions. This allows the hybrid model to capture both the low- and high-level features of the data, which leads to improved performance. From a mathematical formulation perspective, stacking ensemble models are based on the concept of a linear combination of predictions from multiple models. The predictions from each base model are combined using a linear function, and the weights of this function are

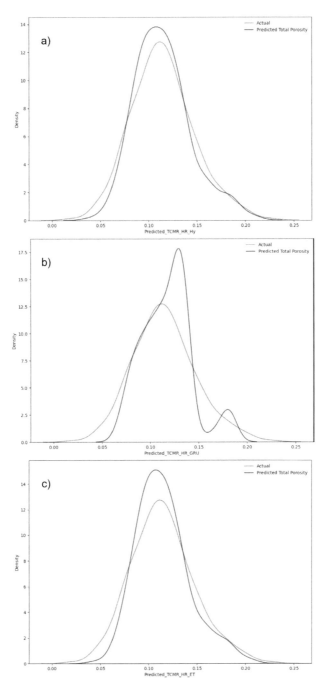

Fig. 3.6. KDE indicating the closeness of estimated to actual total porosity where (a) is hybrid model prediction, (b) is GRU model prediction, and (c) is ET model prediction.

Application of a Novel Stacked Ensemble Model in Predicting Total Porosity   51

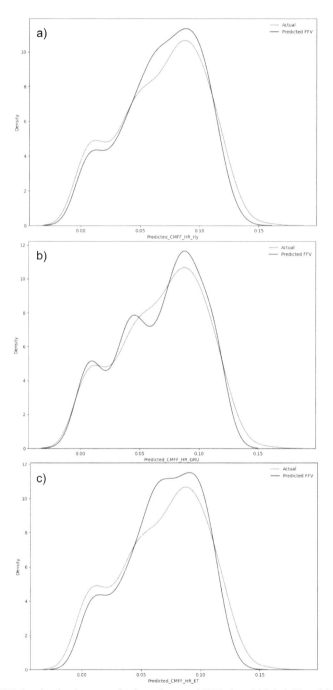

Fig. 3.7. KDE showing the closeness of estimated to actual FFV where (a) is hybrid model prediction, (b) is GRU model prediction, and (c) is ET model prediction.

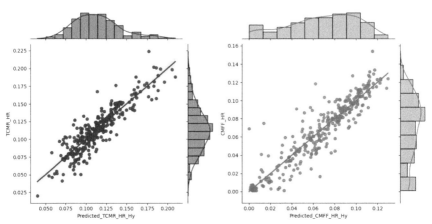

Fig. 3.8. Joint plot of the hybrid model predicted vs actual values where total porosity is on the left and FFV is on the right.

learned by training a meta-model on the predictions from the base models. This combination of predictions from multiple models results in a more robust model that generalises well to new data, as exhibited by the results. This establishes the foundation of its superior prediction performance over the independent base learners.

It is important to note that the LSTM, CNN, and GRU models performed poorly, which may be due to the limited amount of data or the complexity of the problem. Training a DL model on small datasets has the propensity to initiate data memorisation, leading to poor performance on the holdout data. The ineffectiveness of DL involving small datasets and its incapability to interpret results confirms that it cannot be excessively adjudged the aptest solution in all cases

Training a DL model on small datasets causes data memorisation, often leading to overfitting, resulting in weak performance on the holdout data. The limited effectiveness of DL on small data and its incapability to interpret results suggest that it cannot be solely considered the optimal solution in all scenarios. Overall, the excellent performance of the hybrid model can be attributed to its ability to combine the strengths of multiple base models and its use of a stacking ensemble technique and mathematical formulation that leads to a more robust model that generalises well to new data. The hybrid model was used to predict the total porosity and FFV for the entire well, as shown in Fig. 3.9. The predicted model includes the training and the testing data.

Application of a Novel Stacked Ensemble Model in Predicting Total Porosity 53

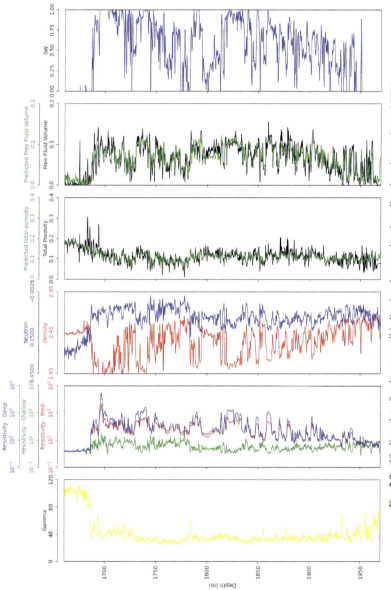

Fig. 3.9. Wireline plot of carbonate well indicating the actual and predicted total porosity and FFV.

5. Conclusions

To obtain a detailed understanding of the reservoir, it is crucial to quantify the surface wettability at an early stage. This information is essential for enhanced oil recovery (EOR), particularly chemical EOR, to maximise production. This study aimed to develop a suitable model for predicting total porosity and the free fluid volume using wireline logs as input. The study evaluated several models, including a Hybrid ensemble model, Extra Trees Regression, Random Forest, XGBoost, LSTM, GRU, and CNN. The performance of the models was evaluated using several metrics, including train score, test score, AIC, MAE, and RMSE. The incorporation of these computational methods provides a more accurate depiction of surface wettability when core wettability data and NMR logs are unavailable.

In the light of the results obtained in this study, the following inferences can be made;

1. The hybrid ensemble model achieved a high train and test accuracy score, indicating that it is able to fit the training data well and generalise well to new data. This is further supported by the recorded lowest MAE, and RMSE in its predictions.
2. The likelihood of the model fitting new data is also high, as indicated by the AIC score, which is the lowest among all the models. AIC is a measure of the probability of the model's fit on new data. A low AIC score indicates that the model is more likely to generalise well to new data.
3. The DL models, LSTM, CNN, and GRU, performed poorly due to their high complexity and volume of data used.
4. The hybrid ensemble model performed better than the others because it combined the strengths of multiple base models, Extra Trees Regression, Random Forest, and XGBoost, and used a stacking ensemble technique. This technique involves training multiple base models on the input data and then using their predictions as input to a higher-level meta-model that makes the final predictions. This allows the hybrid model to capture both the low-level and high-level features of the data, which leads to improved performance.

The results of this study suggest that the hybrid ensemble model is a robust fit for this problem and is suitable for predicting total porosity and FFV. It is able to achieve high accuracy, consistency, and precision, and generalise well to new data. This makes it a valuable tool for the petroleum industry in estimating the porosity of reservoirs, determining the pore volume fraction of a reservoir that can be produced, and evaluating the potential of a prospect. This can help improve hydrocarbon exploration and production efficiency, ultimately leading to increased productivity and reduced costs.

Acknowledgement

The authors express their sincere appreciation to Universiti Teknologi Petronas and the Centre of Research for Subsurface Seismic Imaging for supporting this work.

References

Amani, M., Al-Jubouri, M.B., Khadr, S. and Sayed, A.M. 2017. A comprehensive review on the use of NMR technology in formation evaluation. https://www.semanticscholar.org/paper/A-Comprehensive-Review-on-the-Use-of-NMR-Technology-Amani-Al-Jubouri/fe15a754ea33ae2a3c1f2fa689b81f067b398fef.

Bloch, F. 1946. Nuclear induction. *Physical Review* 70: 460. https://doi.org/10.1103/PhysRev.70.460.

Branco, F.R. and Gil, N.A. 2017. NMR study of carbonates wettability. *J. Pet. Sci. Eng.* 157: 288–294. https://doi.org/10.1016/j.petrol.2017.06.023.

Cai, J. and Hu, X. 2019. *Petrophysical Characterization and Fluids Transport in Unconventional Reservoirs.* pp. 1–332. https://doi.org/10.1016/C2018-0-00934-2.

Coates, G.R., Xiao, L. and Prammer, M.G. 1999. *NMR Logging Principles and Applications.* Houston: Halliburton Energy Services Publication.

Dugan, B. 2015. Data report: Porosity and pore size characteristics of sediments from Site C0002 of the Nankai Trough determined by mercury injection. https://doi.org/10.2204/IODP.PROC.338.202.2015.

Elsayed, M., Glatz, G., El-Husseiny, A., Alqubalee, A., Adebayo, A., Al-Garadi, K. and Mahmoud, M. 2020. The effect of clay content on the spin–spin NMR relaxation time measured in porous media. *ACS Omega* 5: 6545–6555. https://doi.org/10.1021/acsomega.9b04228.

Gu, M., Xie, R. and Jin, G. 2021. A machine-learning based quantitative evaluation of the fluid components on T2-D spectrum. *Mar. Pet. Geol.* 134: 105353. https://doi.org/10.1016/J.MARPETGEO.2021.105353.

Gubelin, G. and Boyd, A. 1997. Total porosity and bound-fluid measurements from an NMR tool. *Journal of Petroleum Technology* 49: 718–718. https://doi.org/10.2118/39096-JPT.

Heaton, N., Minh, C.C., Freedman, R. and Flaum, C. 2000. High-resolution bound-fluid, free-fluid and total porosity with fast NMR logging. *Paper presented at the SPWLA 41st Annual Logging Symposium, Dallas Texas (June).* https://www.slideshare net>armhaggag>high resolute…

Li, C., Liu, X., You, F., Wang, P., Feng, X. and Hu, Z. 2022. Pore size distribution characterization by joint interpretation of MICP and NMR: A case study of chang 7 tight sandstone in the ordos basin. *Processes* 10: 1941. https://doi.org/10.3390/pr10101941.

Li, H., Misra, S. and He, J. 2020. Neural network modeling of *in situ* fluid-filled pore size distributions in subsurface shale reservoirs under data constraints. *Neural Comput. Appl.* 32: 3873–3885. https://doi.org/10.1007/S00521-019-04124-W.

Masroor, M., Emami Niri, M., Rajabi-Ghozloo, A.H., Sharifinasab, M.H. and Sajjadi, M. 2022. Application of machine and deep learning techniques to estimate NMR-derived permeability from conventional well logs and artificial 2D feature maps. *Journal of Petroleum Exploration and Production Technology* 12(3): 2937–2953. https://doi.org/10.1007/S13202-022-01492-3.

Mitchell, P., Al-Hosani, I., Mehairi, Y. Al and Kalam, M.Z. 2008. Importance of mercury injectiion capillary pressure (MICP) measurements at pseudo reservoir conditions.

Society of Petroleum Engineers –13th Abu Dhabi International Petroleum Exhibition and Conference. ADIPEC 2008 2: 952–962. https://doi.org/10.2118/117945-MS.
Otchere, D.A., Arbi Ganat, T.O., Gholami, R. and Ridha, S. 2021a. Application of supervised machine learning paradigms in the prediction of petroleum reservoir properties: Comparative analysis of ANN and SVM models. *J. Pet. Sci. Eng.* 200: 108–182. https://doi.org/10.1016/j.petrol.2020.108182.
Otchere, D.A., Ganat, T.O.A., Ojero, J.O., Taki, M.Y. and Tackie-Otoo, B.N. 2021b. Application of gradient boosting regression model for the evaluation of feature selection techniques in improving reservoir characterisation predictions. *J. Pet. Sci. Eng.* 109244. https://doi.org/10.1016/J.Petrol.2021.109244.
Otchere, D.A., Hodgetts, D., Ganat, T.A.O., Ullah, N. and Rashid, A. 2021c. Static reservoir modeling comparing inverse distance weighting to kriging interpolation algorithm in volumetric estimation. case study: gullfaks field. *In: Offshore Technology Conference.* OnePetro, Virtual and Houston, Texas. https://doi.org/10.4043/30919-MS.
Otchere, D.A., Abdalla Ayoub Mohammed, M., Ganat, T.O.A., Gholami, R. and Aljunid Merican, Z.M. 2022a. A novel empirical and deep ensemble super learning approach in predicting reservoir wettability via well logs. *Applied Sciences* 12: 2942. https://doi.org/10.3390/app12062942.
Otchere, D.A., Ganat, T.O.A., Nta, V., Brantson, E.T. and Sharma, T. 2022b. Data analytics and Bayesian Optimised Extreme Gradient Boosting approach to estimate cut-offs from wireline logs for net reservoir and pay classification. *Appl. Soft. Comput.* 120: 108680. https://doi.org/10.1016/j.asoc.2022.108680.
Peng, S., Zhang, T., Loucks, R.G. and Shultz, J. 2017. Application of mercury injection capillary pressure to mudrocks: Conformance and compression corrections. *Mar. Pet. Geol.* 88: 30–40. https://doi.org/10.1016/J.MARPETGEO.2017.08.006.
Purcell, W.R. 1949. Capillary pressures –their measurement using mercury and the calculation of permeability therefrom. *Journal of Petroleum Technology* 1: 39–48. https://doi.org/10.2118/949039-G.
Rezaee, R. 2022. Synthesizing Nuclear Magnetic Resonance (NMR) outputs for clastic rocks using machine learning methods, examples from north west shelf and perth basin, Western Australia. *Energies* (Basel) 15: 518. https://doi.org/10.3390/en15020518.
Tamoto, H., Gioria, R. dos S. and Carneiro, de C. 2023. Prediction of nuclear magnetic resonance porosity well-logs in a carbonate reservoir using supervised machine learning models. *J. Pet. Sci. Eng.* 220: 111169. https://doi.org/10.1016/j.petrol.2022.111169.
Tarafder, S., Badruddin, N., Yahya, N. and Nasution, A.H. 2022. Drowsiness detection using ocular indices from EEG signal. *Sensors* 22: 4764. https://doi.org/10.3390/s22134764.
Wu, B., Xie, R., Wang, X., Wang, T. and Yue, W. 2020. Characterization of pore structure of tight sandstone reservoirs based on fractal analysis of NMR echo data. *J. Nat. Gas Sci. Eng.* 81: 103483. https://doi.org/10.1016/j.jngse.2020.103483.
Wu, B., Xie, R., Xu, C., Wei, H., Wang, S. and Liu, J. 2021. A new method for predicting capillary pressure curves based on NMR echo data: Sandstone as an example. *J. Pet. Sci. Eng.* 202: 108581. https://doi.org/10.1016/j.petrol.2021.108581.
Xu, H., Tang, D., Zhao, J. and Li, S. 2015. A precise measurement method for shale porosity with low-field nuclear magnetic resonance: A case study of the Carboniferous–Permian strata in the Linxing area, eastern Ordos Basin, China. *Fuel* 143: 47–54. https://doi.org/10.1016/j.fuel.2014.11.034.

CHAPTER 4

Compressional and Shear Sonic Log Determination
Using Data-Driven Machine Learning Techniques

Daniel Asante Otchere,[1,2,*] *Raoof Gholami*,[3] *Vanessa Nta*[4] and *Tarek Omar Arbi Ganat*[5]

1. Introduction

Shear (Vs) and Compressional (Vp) sonic waves are significant parameters in subsurface engineering. They provide valuable information for reservoir exploration, development, recovery and fluid sequestration (Azizia et al., 2017; Zoveidavianpoor et al., 2013). These parameters or its ratio are useful in reflection seismology, lithologic identification, and formation evaluation (Castagna et al., 1985), pore fluid and pore pressure information (Duffaut and Landrø, 2007; Rojas, 2008), reservoir characterisation (Eberli et al., 2003; Pickett, 1963), geophysics (Phadke et al., 2000; Waluyo et al., 1995), and geomechanical properties (Asoodeh and Bagherıpour, 2014; Rasouli et al., 2011). There are several ways of measuring these parameters, as summarised in Table 4.1.

---

[1] Centre of Research for Subsurface Seismic Imaging, Universiti Teknologi PETRONAS, 32610, Seri Iskandar, Perak Daril Ridzuan, Malaysia.
[2] Institute for Computational and Data Sciences, Pennsylvania State University, University Park, PA, USA.
[3] Department of Energy Resources, University of Stavanger, Kjell Arholms gate 41, Stavanger, 4021, Norway.
[4] Shell Oil Company, 150 N Dairy Ashford Rd, Houston, TX, 77079, United States of America.
[5] Department of Petroleum and Chemical Engineering, Sultan Qaboos University, Muscat, Oman.
* Corresponding author: ascotjnr@yahoo.com

Table 4.1. Common methods of estimating Vs sonic velocity.

Methods	Types of Measurement
Laboratory	Core measurements
Direct	Dipole Sonic Log
Empirical Correlations	Castagna, Gassman, Brocher, Carroll and Greenberg–Castagna correlations
Machine Learning	Examples of models

The acquisition of continuous Vs and Vp subsurface data is reliably done via wireline logs. The Dipole Shear Sonic Log, invented in the 1980s, is the most commonly used logging tool used to record the waveform values and process the data to compute Vs and Vp sonic velocity and became ubiquitous in the 1990s (Close et al., 2009). This measuring tool indicates how fast sound propagates through subsurface interfaces. Although the dipole sonic log is available, its expensiveness has rendered the direct measurement of Vs not worth the cost, as such, Vs is not ubiquitous in all wells. This limitation has necessitated further research in estimating Vs from other logs, mainly from the Vp log (Brocher, 2005; Carroll, 1969; Castagna et al., 1985; Dvorkin and Mavko, 2014; Gassmann, 1951; Greenberg and Castagna, 1992). With the advancement of fluid sequestration, geothermal energy, and unconventional resources, Vs and Vp have become more prevalent. However, older wells never acquired Vs and Vp data before the dipole sonic log. Acquiring the data now is not feasible. As such, extensive research is required to provide substantial study in estimating these two logs from other conventional wireline logs. In this study, estimating Vs and Vp from wireline logs using a multioutput supervised machine learning approach is investigated for the Volve field siliciclastic oil-bearing reservoirs. This study is of the utmost significance, especially in more mature fields, as shear and compressional sonic logs were never obtained in old wells. This method would result in better accuracy in predicting compressional and shear sonic logs for hundreds of wells useful for geomechanics, performing well-to-seismic inversion, and other operations.

2. Literature Review

Prior to the influx of data, empirical and laboratory techniques have been useful in estimating Vs from Vp. One of the pioneering articles that were published by (Pickett, 1963) established the concept of using the compressional to shear wave velocities ratio as a lithology indicator. Figure 4.1, which has been modified from Pickett's study, demonstrates the clear distinction in Vp/V that exists between dolomites, limestones, and clean sandstones. According to (Castagna et al., 1985), the Vp/Vs ratio tends to change fairly linearly between the velocity ratios of the end members with increasing composition in binary

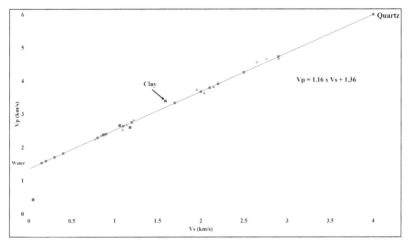

Fig. 4.1. Vs and Vp sonic velocities from *in-situ* sonic and field seismic measurements for mudrocks. Modified after Castagna et al. (1985).

combinations of quartz and carbonates. The results of this study served as the foundation of several empirical approaches. This study will discuss two of the most commonly used techniques for estimating Vs.

Research conducted by (Castagna et al., 1985) concluded that Vs sonic velocity is directly related to Vp sonic velocity for both water-saturated and dry siliciclastic sedimentary rocks, as shown in Fig. 4.1. Shales tend to have a relatively greater Vp/Vs than clean sandstones, given a similar Vp. From their results, Vp/Vs is almost homogeneous for dry sandstones. Wet sandstones and mudstones exhibited an indirect monotonic relationship between Vp and Vp/Vs. The Vs sonic velocities measured in water-saturated sandstone agreed with those predicted by Gassmann's equations. This empirical linear relationship between the Vp and Vs sonic velocity in brine-saturated clastic silicate rocks is known as the mudrock line and is expressed from *in-situ* sonic and field seismic measurements as:

$$V_p = 1.16 V_s + 1.36 \qquad (4.1)$$

This linear equation is partly explained by the location of the clay point near a line joining the quartz point with the water velocity. The equation was formulated assuming that Vp and Vs velocities decrease linearly as increasing clay porosity approaches water. Similarly, when pure clay is mixed with quartz, velocities increase monotonically as the quartz point is approached. These constraints correspond with those deduced from Tosaya and Nur (1982) empirical relations, except for high porosity behaviour.

Greenberg and Castagna (1992) proposed a correlation, the Greenberg-Castagna formulation, for pure unit completely brine-saturated mineral rocks based on linear relations of Vp to estimate the Vs sonic velocity from other petrophysical parameters. To predict Vs sonic velocity of a 100% saturated rock, it is required to perform fluid replacement calculations by utilising Gassman relations. In order to mitigate the major limitation of the Greenberg-Castagna formula, fluid substitution is done when there are hydrocarbon-bearing rocks. It requires several sets of operations known as Gassmann fluid substitutions (Gassmann, 1951). They utilised the regression weights for different lithologies shown in Table 4.2 to satisfy the expression:

$$V_s = a_{i2} V_p^2 + a_{i1} V_p + a_{i0} \quad (4.2)$$

The lines of fit for the lithological regression weights are based on Simm and Bacon (2014). The method's fundamental process entails completing multiple iterations beginning with Vp assumptions for brine until it reaches a considerable convergence, assuring the validity of the Vs prediction. The regression coefficients of Vs are presented in Table 4.2.

Techniques based on artificial intelligence (AI) are currently being utilised on a massive scale in the renewable and oil and gas industry. These techniques have led to significant developments in their application to improve the accuracy of their predictions (Otchere et al., 2021b). Numerous research has used machine learning techniques to predict Vs, highlighting their usefulness in identifying patterns amongst highly dimensional and complex wireline data (Anemangely et al., 2019; Bagheripour et al., 2015; Onalo et al., 2018). Table 4.3 summarises the various machine learning research conducted to predict Vs and, in some cases, Vp from conventional wireline logs. The summary will indicate the differences in this current study.

The fundamental differences between the various studies and techniques used in predicting Vs or Vp were the model type, formation, reservoir fluid and input variables. Whereas supervised machine learning models are only as efficient as the information they are trained with when irrelevant data is used as input or important features is omitted, the model's performance degrades (Otchere et al., 2021c). As a result, identifying and selecting features based on a demonstrated record of their effect on Vs and Vp measurements is required, invalidating the choice of machine learning models used as relevant. The main distinction between this study and earlier studies is the multi-output prediction of Vp and Vs, with Vp not being used as an input for Vs prediction.

Table 4.2. Regression coefficients of Vs presented by Castagna et al. (1993).

Lithology	ai2	ai1	ai0	R^2
Sandstone	0	0.80416	−0.85588	0.98352
Limestone	−0.05508	1.01677	−1.03049	0.99096
Dolomite	0	0.58321	−0.07775	0.87444
Shale	0	0.76969	−0.86735	0.97939

Table 4.3. Summary of previous research applying machine learning in predicting Vs and Vp.

Author	Model Used	Lithology	Input Variables	Output Variable(s)	Study Area
Bagheripour et al. (2015)	Support Vector Regression	Carbonate	• Compressional sonic • Density • Neutron • True resistivity • Shallow resistivity • Photoelectric factor • Gamma ray	Shear sonic	Iran Gas fields
Bukar et al. (2019)	Exponential Gaussian Process Regression Model	Sandstone	• Compressional sonic • Caliper • Bulk density correction • True resistivity • Neutron • Porosity • Density • Water saturation • Gamma ray	Shear sonic	Australia Gas fields
Anemangely et al. (2019)	Least Square Support Vector Machine-Cuckoo Optimisation Algorithm	Carbonate	• Compressional sonic • True resistivity • Neutron • Density • Gamma ray	Shear sonic	Iran's Ahvaz and Ab-Teymour oilfields
Nourafkan and Kadkhodaie-Ilkhchi (2015)	Hybrid ant colony–fuzzy inference system (ACOFIS)	Carbonate	• Compressional sonic • Neutron • Density • Gamma ray • Photoelectric factor	Shear wave velocity	Iran Cheshmeh-Khosh oilfield
Elkatatny et al. (2018)	• Adaptive neuro-fuzzy inference system (ANFIS) • Support vector machine (SVM) • Artificial Neural Network (ANN)	Not stated	• Bulk density • Neutron porosity • Gamma ray	• Shear wave • Compressional wave	Not stated

Table 4.3 contd. ...

...Table 4.3 contd.

Author	Model Used	Lithology	Input Variables	Output Variable(s)	Study Area
Gamal et al. (2022)	• Decision Tree (DT) • Random Forest (RF)	• Limestone • Sandstone • Shale	surface drilling parameters	• Shear slowness • Compressional slowness	Not stated
LeCompte et al. (2021)	Artificial neural network (ANN)	Sandstone	• Drilling diagnostics (EDR) data • Resistivity log • Gamma ray	• Compressional sonic • Density	Gulf of Mexico
Suleymanov et al. (2021)	• Adaptive neuro-fuzzy inference system (ANFIS) • Support vector machine (SVM)	Not stated	• Weight on bit (WOB) • Rate of penetration (ROP) • Standpipe pressure (SPP) • Torque (T) • Drill pipe rotation (RPM) • Mud flow rate (GPM)	• Shear wave slowness • Compressional wave slowness	Not stated
Ali et al. (2021)	Deep Neural Network (DNN)	• Shale • Sandstone	• Sonic log • Deep resistivity • Neutron • Depth • Gamma ray	Shear sonic log	Lower Goru Formation in the Middle Indus Basin, Pakistan
Asoodeh and Bagheripour (2012)	• Artificial neural network • Fuzzy logic algorithm • Neuro-fuzzy algorithm	Carbonate	• Thermal neutron porosity (NPHI) • Bulk density (RHOB) • Electrical resistivity (Rt) • Shale volume (Vsh)	• Compressional wave velocity • Shear wave velocity • Stoneley wave velocity	Asmari formation

Safaei-Farouji et al. (2022)	• Random Forest (RF) • Extra Tree (ET) • Gaussian Process Regression (GPR) • Integration of Adaptive Neuro Fuzzy Inference System (ANFIS) with differential evolution (DE) and imperialist competitive algorithm (ICA) optimisers	Carbonate	• Compressional sonic wave velocity • Neutron • Density • Gamma ray	Shear sonic wave velocity	Sarvak oilfield, Iran
Hatampour and Ghiasi-Freez (2013)	Fuzzy Logic	Carbonate	• Neutron • Density • Electrical resistivity (Rt)	• Compressional wave velocity • Shear wave velocity • Stoneley wave velocity	Iran
Chaikine and Gates (2020)	Hybrid convolutional-recurrent neural network (cRNN)	• Siltstone • Shale	• Well deviation survey • Density • Compressional slowness	Shear slowness	Montney Formation in Alberta, Canada
Zoveidavianpoor et al. (2013)	Adaptive neuro-fuzzy inference system	Carbonate	• Neutron • Density • Gamma ray	Compressional wave	Middle Eastern Oilfield

Data from three wells in the Volve oil field is used in this study. The input features and their relevance will be evaluated using the laboratory-verified justification of their influence on Vs and Vp prediction. After establishing a suitable dataset of relevant input variables, different machine learning models will be employed to predict Vp and Vs. The best-performing model based on several statistical metrics will then be optimised using the Bayesian Optimisation (BO) algorithm.

3. Background of Machine Learning Regression Models

This study builds on previous research in which multiple procedures and strategies were utilised to forecast Vs or Vp utilising available input information. Machine learning is becoming increasingly popular as academics move away from empirical correlations. Individual well log parameters provide crucial information about Vp and Vs. However, when particular subsets are employed, the outcomes may be estimated more precisely. In some instances, researchers use the filter method to select critical features, while others use wrapper methods, intrinsic techniques, and metaheuristics algorithms to select relevant input features. This shows that there is no standard set of features for this prediction. These strategies, however, are based on statistical measurements rather than domain knowledge. As such, understanding Vp and Vs and how subsurface parameters influence their measurements is utilised to solve this prediction problem. The appropriate features selected will subsequently be used as input features. Several machine learning models based on Decision Tree learning will be used to predict Vp and Vs.

3.1 Decision Tree Conceptual Overview

The Decision Tree algorithm has been around for decades, but it is currently known as Classification and Regression Trees (CART). CART, in contrast to other supervised learning models, can be used to address regression and classification problems (Tarafder et al., 2022). The aim of utilising a Decision Tree is to build a training model to estimate the value or class of an output variable by learning basic decision rules using prior data (Breiman, 2001). The CART model is the basis for major machine learning models like the Random Forests, bagged decision trees, and boosted decision trees. The CART model is a binary tree similar to algorithms and data structures, with no, single or two child nodes for each node. Assuming the parameter is numeric, a node defines a specific input parameter (X) and a split point on that parameter. The tree's terminal (leaf) nodes include a target parameter (y) utilised to make a decision. Upon building the tree, the tree can be traversed by inserting a new row of data after each branch with splits until the final prediction. Making binary decision trees is a technique for splitting up the input data. A greedy technique known as recursive binary splitting is utilised for the data partition. Recursive binary splitting is a statistical approach in which all values are aligned, and several strategic split points are explored and assessed using an objective function. The split that results in the lowest cost is thence chosen. Based on the objective function, all input parameters and feasible splits are assessed and selected in a greedy approach.

The Disjunctive Normal Form concept, also known as the Sum of Products (SOP), is the foundation of the CART model. Various branches that end in the same class combine to form a sum for each branch that extends from the tree's root to a leaf node of that class. In implementing the decision tree, determining which qualities should be regarded as the root node at each level is the primary issue. Addressing this is referred to as attribute selection. Different attribute selection techniques determine the attribute designated as the root note at each level.

The preference to do judicious splits has a major impact on tree's reliability. There are specific selection principles for CART models. To determine whether a node should be split into multiple sub-nodes, different algorithms are used by CART models. The formation of sub-nodes improves the uniformity of newly emerging sub-nodes (Gordon et al., 1984). The decision to do strategic splits heavily affects a tree's accuracy, meaning that the integrity of the node grows with the target variable. The CART model divides the nodes based on all available parameters and then chooses the split that produces the most homogenous subnodes. The algorithm selection is also chosen depending on the type of target data.

3.1.1 Attribute Selection Measures

Considering a dataset has N parameters, selecting which ones to put at the root or various levels of the tree as internal nodes is a complicated process. Choosing any node at random to be the root will not address the problem. Choosing a random approach may result in bad outcomes with low precision. Researchers worked on this attribute selection issue and proposed several approaches utilising parameters such as (Breiman, 2001):

1. *Entropy (E)*: The measure of how random information is being processed is called entropy. High entropy makes it difficult for inferences to be made from the data. A branch with zero entropy is a leaf node, while a branch with a zero entropy needs to be split further. Entropy for a parameter is expressed mathematically as:

$$E(S) = \sum_{i=1}^{n} -p_i \log_2 p_i \qquad (4.3)$$

where S denotes the present state, and Pi is the probability of an event i occurring in a node of state S. Entropy for multiple parameters is expressed mathematically as:

$$E(S,Y) = \sum_{n \in Y} P(n)E(n) \qquad (4.4)$$

where Y represents the selected parameter.

2. *Information Gain (IG)*: IG is a numerical attribute that quantifies how efficiently a specific attribute distinguishes training samples according to their target. Building a decision tree is to identify a characteristic with the most considerable information gain and the lowest entropy. Entropy decreases as a result of information gain. Based on the specific attribute values provided, it calculates the difference between the data's entropy before splitting and the average entropy after splitting. IG is denoted mathematically as:

$$IG(S,Y) = E(S) - E(S,Y) \qquad (4.5)$$

3. *Gini Index (GI)*: The Gini index may be considered an objective function to assess splits in a dataset. It is computed by subtracting the total of each class's squared probability from one. GI prefers large divisions that are simple to implement, whereas information gain prefers fewer divisions with unique values. The Gini Index only performs binary splits by employing the binary target variable 'Success' or 'Failure'. A high GI value indicates greater inequality and heterogeneity. GI can be mathematically denoted as:

$$GI = 1 - \sum_{i=1}^{n} -(p_i)^2 \qquad (4.6)$$

4. *Gain Ratio*: IG is biased towards choosing attributes with many values as root nodes. It means it prefers the attribute with a large number of distinct values. Gain ratio, a variation of IG that minimises bias, is typically the best option compared to IG. The gain ratio solves the issues related to IG by considering the number of branches that might occur before splitting. It adjusts IG by considering the inherent information of a split. The gain ratio is expressed as:

$$GI = \frac{E(S) - E(S,Y)}{\sum_{j=0}^{K} w_j \log_2 w_j} \qquad (4.7)$$

where K refers to the number of subsets produced by the split, and j, refers to a subset.

5. *Reduction in Variance (RIV)*: RIV is a technique used for regression problems. This method uses the traditional variance formula to select the optimum split. The split achieving the lowest variance is chosen as the criterion for splitting the data:

$$Variance = \frac{\sum(X - \bar{X})^2}{n} \qquad (4.8)$$

The mean of the values is shown as $\bar{X}$. X is the actual value, and n is the number of values.

6. *Chi-Squared Automatic Interaction Detector (CHAID)*: CHAID is a tree classification approach that determines the statistically significant differences between sub-nodes and the parent node. The sum of squares of the standardised discrepancies between the observed and predicted frequencies of the target is used to calculate it. CHAID employs the binary target variable 'Success' or 'Failure' by performing two or more splits. The greater the Chi-Square value, the greater the statistical significance of discrepancies between the sub-node and Parent node. Chi-squared is denoted mathematically as:

$$x^2 = \sum \frac{(O - E)^2}{E} \qquad (4.9)$$

where x^2 represents the obtained Chi-Square, O is the observed score, and E denotes the expected score.

These criteria will be used to calculate each attribute selection value. Each value is ranked, while the characteristics are orderly put in the tree. The attribute with the highest value regarding knowledge gain is at the root. The categorical output is assumed when utilising IG as a criterion, whereas GI assumes continuous attributes. CART models are simple to understand since they result in precepts, but one main disadvantage is the practical possibility of overfitting. Overfitting generally happens when CART builds many branches due to outliers and irregularities in data, but it can be minimised using the pre- and post-pruning approaches. Pre-pruning terminates tree construction early, and it is preferable not to split a node if its measure of purity falls below a certain threshold. However, deciding on an acceptable stopping point is challenging. Post-pruning begins by going deeper into the tree to construct an entire tree. If the tree exhibits overfitting, pruning is performed as a post-pruning phase. Cross-validation is employed to determine if extending a node would improve or not to evaluate the efficacy of pruning. If there is an improvement, extending that node can be done. However, if it indicates a degraded performance, the node should not be extended and changed to a terminal node (Breiman, 2001; Gordon et al., 1984).

3.2 Random Forest Conceptual Overview

Another well-known robust bagging algorithm is the Random Forest (RF) developed by Breiman (Breiman, 2001). RF ensembles numerous Decision Tree base learners randomly using averaging (Dinov, 2018). Each base learner is trained on a random vector independently chosen with similar vector distribution to the other forest learners. The wisdom of crowds is a simple yet powerful notion at the core of RF. This concept means that many relatively uncorrelated learners working together surpass any individual base learners. The crucial point is the poor correlation amongst models, which allows the excellent performance of RF. The trees shield each other from their prediction errors, so RF generates excellent outcomes. While some trees may be inaccurate, many others will be accurate, allowing the trees to grow in the appropriate direction as a forest. Decision trees are susceptible to the dataset on which they are trained such that minor changes to the training data typically result in a radically altered tree structure. This disadvantage is solved by the random forest using the bagging principle, which allows each unique tree to be randomly selected from the dataset with replacement, resulting in diverse trees.

Any given tree's prediction becomes, in a sense, a characteristic of the overall model prediction. Due to its robustness against overfitting and simple interpretability, RF is frequently employed to provide good model results. This merit results from its assessment of the target's critical input factors and the high priority given to correlated aspects. However, RF has a greater bias against high cardinal characteristics. The averaging of the base learners' predictions is mathematically represented as:

$$\hat{y} = \frac{1}{n}\sum_{k=1}^{n} h_k(X) \quad (4.10)$$

where $\hat{y}$ is the output, h(X) is the number of trees, X is the input vector, and n and k are the overall numbers of trees grown where n is greater than k and k is greater than 1.

The out-of-bag (OOB) score is another essential aspect of the RF algorithm. Specific samples will be excluded from the subsamples used to train the base learners when bootstrapping. These out-of-sample examples may be used to assess the learner and generate an OOB score, acting as a pseudo-validation subset for the random forest model. Set oob score = True when initialising the random forest object to acquire the OOB score. Another critical aspect of the RF algorithm is its feature engineering capability entailing selecting the most relevant features from the input variables in the training dataset. Feature selection is an essential part of the machine learning workflow.

The primary disadvantage of random forest is that many trees may slow down and render the model inefficient for predictions. In general, the model tends to learn rapidly but makes slow predictions. A more precise prediction necessitates more trees, resulting in a computationally exhaustive model.

3.3 Extremely Randomised Trees Conceptual Overview

Extremely Randomised Trees (Extra-Trees) is an ensemble learning technique proposed by Geurts et al. (Geurts et al., 2006). Tree-based regressors are composed of a hierarchy set of rules that may predict the numerical values of the output (Gordon et al., 1984). The Extra-Trees technique uses the traditional top-down strategy to construct an ensemble of unpruned decision trees. The two key distinctions from previous tree-based ensemble algorithms are that it separates nodes by selecting cut-points entirely at random. It grows the trees using the entire training sample instead of replicating a bootstrap replica. Extra-Trees appropriately average the randomised predictions of Decision Trees to enhance precision while minimising computational complexity significantly.

The Extra-Trees approach is built on the bias-variance concept. The explicit randomisation of the cut-point and features combined with ensemble averaging can reduce variance more strongly than the weaker randomisation techniques employed by other algorithms. The entire original training data should be used to reduce bias rather than bootstrap replicas. The computational cost of the tree growing method, assuming balanced trees, is on the order of N log N regarding the learning sample size, as is the case with most other tree-growing procedures. Moreover, given the ease of the node splitting technique, it is anticipated that the constant factor will be considerably lower than in previous ensemble-based methods that optimise cut points locally (Geurts et al., 2006).

Constructing the nodes and branches that make up a tree depends on dividing the input vector into independently exclusive areas based on a predetermined splitting criterion, gradually reducing the size of the regions. When the number of samples in a tree section goes below a predefined threshold, the splitting terminates to form a leaf. When inputting a new sample into the tree, a specific path is taken based on the splitting criteria provided in the tree-building process. The predicted target is generated by aggregating the values stored in the leaf. The requirements for splitting, termination into leaves, tree quantity developed, and assigning weight to each leaf are the critical aspects that distinguish Extra-Trees from other tree-based techniques.

Aside from having similar advantages as single trees based on consistency for universal generalisation and approximation, Extra-Trees also give resilience in gross model errors because outliers affect its predictions slightly and locally. Extra-Trees potentially outperform single trees in relation to computational efficiency while resisting irrelevant input features when the input features are significantly more than the random splits. Further comparison with single trees and dependent on the issue at hand indicate that

Extra-Trees exhibit low variance, and their prediction accuracy improves with an increasing number of trees (Ernst et al., 2006).

4. Data Collection and Description

The Volve field, located in the Norwegian part of the North Sea, is a sandstone oil-bearing reservoir interval. The Volve field formation predominantly consists of near-shore, shallow, to marginal marine deposits. As a result of the extensive core sampling, it was discovered that six distinct facies associations represent the Heather Formation; coastal plain, shoreface, fluvio-tidal channel-fill, bay-fill, mouth bar, and offshore open marine (Folkestad and Satur, 2008; Otchere et al., 2022b, 2021a; Sneider et al., 1995). The data used has 12 dimensions compiled from three wells, Well 15_9-F-1_A (Well A), 15_9-F-1_B (Well B) and 15_9-F-15_D (Well C). However, well A and B were combined, making up 4,000 data points (2,125 for Well A and 1,875 for Well B), to train and test the model. A wireline plot of Wells A and B showing the shear (DTS) and compressional (DT) logs is illustrated in Figs. 4.2 and 4.3. The model was then deployed to estimate DTS and DT for Well C, which had a total data point of 11,935 without Vp and Vs logs. One of the main advantages of applying data analytics to data is to find patterns and hidden information in high dimensional data. As such, the primary input features for this research are summarised and justified in Table 4.4. Table 4.5 shows the descriptive statistics of the features for the combined Well A and Well B.

Table 4.4. Justification for log selection as input variables.

Log	Purpose	Reason
Gamma ray (GR)	Lithology determinant	Lithogies exhibit different Vs and Vp wave velocities (Otchere et al., 2021c).
Neutron (NPHI)	Lithology and porosity determinant	Lithogies and porosity influences Vs and Vp differently (Otchere et al., 2021c).
Density (RHOB)	Lithology and porosity determinant	Lithogies and porosity influences Vs and Vp differently (Otchere et al., 2021c).
Photoelectric factor (PEF)	Determination of minerals for lithology interpretation	The response of Vs and Vp varies with different mineral types or compositions (Castagna et al., 1985).
Porosity (PHIF)	Lithology determinant	Vs velocity is more sensitive than Vp to porosity (Hamada and Joseph, 2020).
Water saturation (SW)	Fluid determinant	Different formation fluids affect Vp and Vs responses differently (Castagna et al., 1985).
Resistivity [true (RT), shallow (RACELM), medium (RACEHM), and deep (RPCELM)]	Lithology and hydrocarbon detection logs	Different fluids have different effects on Vs and Vp (Otchere et al., 2021c).
Caliper (CALI)	Detection of permeable zones hence relates to formation types	Different permeabilities affect Vs and Vp differently (Hamada and Joseph, 2020).

Compressional and Shear Sonic Log Determination 71

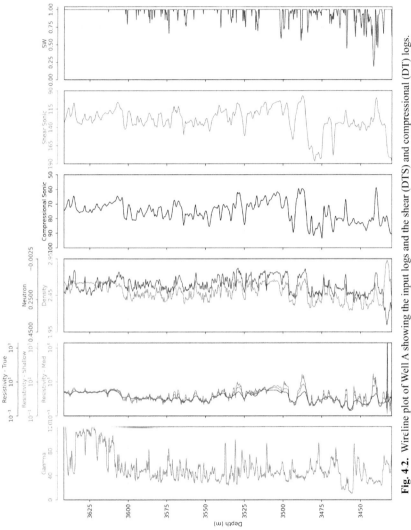

Fig. 4.2. Wireline plot of Well A showing the input logs and the shear (DTS) and compressional (DT) logs.

72 *Data Science and Machine Learning Applications in Subsurface Engineering*

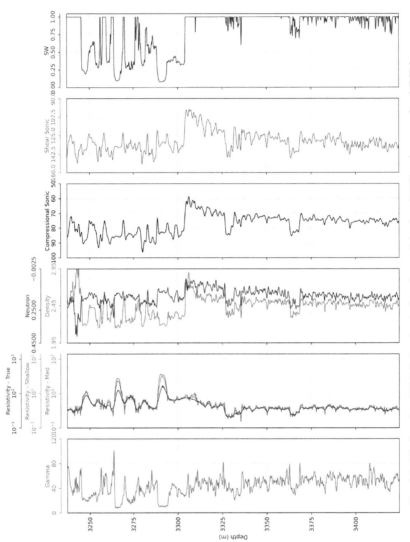

Fig. 4.3. Wireline plot of Well B showing the input logs and the shear (DTS) and compressional (DT) logs.

Table 4.5. Descriptive statistics of the outputs and input variables for the combined Wells A and B.

	CALI	DT	DTS	GR	NPHI	PEF	RACEHM	RACELM	RHOB	RPCELM	RT	PHIF
count	4000	4000	4000	4000	4000	4000	4000	4000	4000	4000	4000	4000
mean	8.66	75.98	129.96	48.75	0.16	6.07	4.25	3.40	2.47	3.02	3.96	0.11
std	0.04	6.80	13.56	20.18	0.05	0.75	87.41	50.11	0.13	7.85	11.99	0.06
min	8.47	58.63	96.90	8.00	0.05	4.30	0.20	0.23	2.15	0.14	0.19	0.02
25%	8.63	71.68	122.19	36.88	0.13	5.62	1.04	1.03	2.40	1.06	1.08	0.06
50%	8.67	75.14	130.42	46.23	0.16	6.05	1.44	1.38	2.48	1.50	1.58	0.10
75%	8.70	80.93	136.00	56.55	0.19	6.37	2.53	2.42	2.56	2.64	3.03	0.15
max	8.87	96.22	186.09	127.06	0.41	10.99	5464.37	2189.60	3.05	96.43	134.70	0.27

5. Methodology

5.1 Data Analysis and Visualisation

The data for this study were analysed and visualised to help understand how the input variables correlate to both targets. The Kendall correlation covariance heatmap was used to calculate the degree of correlation between the inputs and the targets. The NPHI, RHOB, and PHIF were the parameters that strongly correlated to both targets. The nonlinear distribution between the input and the targets is also illustrated in the pair plot in Fig. 4.4. Based on the nonlinear distribution, it is clear that none of the input parameters had a linear relationship to both targets, hence advising on the selection of nonlinear models for this study.

5.2 Machine Learning Model Application

As part of the machine learning process, the models are trained on data which were randomly sampled from the total dataset using the holdout cross-validation technique. This technique is used to assess the performance of models and prevents selection bias and overfitting by predicting the targets using unseen data (Otchere et al., 2022a; Tarafder et al., 2021). The holdout approach divided the data into a 90:10 split, making up 3,600 training and 400 testing data points. The selection of the split is to ensure that the majority of the data are used to train the model due to the heterogeneity of the subsurface but to ensure a suitable number unknown to the models is kept for testing. Different multi-output models were utilised in this study to determine which model accurately predicts the Vs and Vp accurately. This study utilised the same out-of-sample data to guarantee that all the models were predicted on equivalent data points.

The performance of the models was assessed, in order of ranking, using the mean absolute error (MAE), root mean squared error (RMSE), mean

74 *Data Science and Machine Learning Applications in Subsurface Engineering*

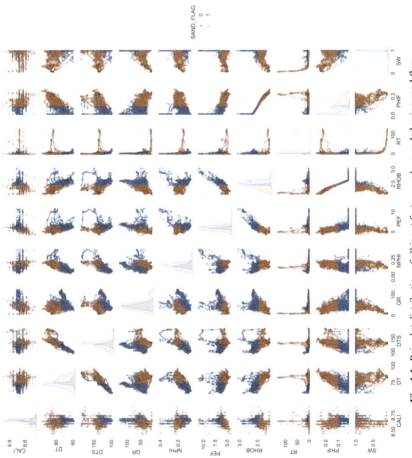

Fig. 4.4. Pair plot distribution of all input features colour-coded using sand flag.

absolute per cent error (MAPE) and R^2. The models were also compared using the Akaike Information Criterion (AIC), a frequentist probability framework that scores models based on the maximum likelihood of fitting unknown data. Upon selecting the best model based on these criteria (BO), hyperparameter tuning, a sequential model-based optimisation procedure, was used to enhance the model's prediction accuracy. The BO technique is a convenient and dynamic paradigm that uses the Bayes Theorem to give a reasonable mechanism for guiding the search of a global optimisation problem to the extrema of objective functions (Otchere et al., 2022b). The technique operates by training a Bayesian approximation probabilistic model of the objective function repeatedly based on previous result estimation. Before choosing acceptable samples for evaluation on the real objective function, the probabilistic model is evaluated using an acquisition function. The process of adjusting the model's parameters to improve the learning algorithm's efficiency and optimise model performance by reaching a sufficiently reduced cost function is known as hyperparameter tuning (Otchere et al., 2022b). The optimised model is then deployed on a new well from the same field to predict Vs and Vp from the wireline logs. The predicted Vs and Vp log will be compared to that obtained from empirical correlations. Figure 4.5 shows the research workflow.

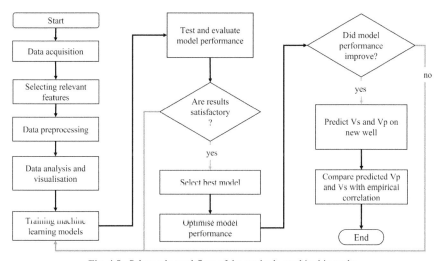

Fig. 4.5. Schematic workflow of the methods used in this study.

6. Results and Discussion

6.1 Evaluation of Model Performance

The models employed in this study were assessed using AIC to identify the model with the highest likelihood of fitting out-of-sample data. The procedure in using this metric to evaluate the models suggests that the lower the AIC values, the greater the possibility of fitting new data. The following considerations were taken for this study when comparing and selecting the best model. There are four levels of significance when it comes to differences in AIC results between models: negligible (< 20), moderate (21–50), substantial (51–100) and extremely strong (> 100). The Extra-Trees model resulted in the highest probability of fit for Vp and Vs predictions by recording the lowest AIC value. When the Extra-Trees model was compared to the Random Forest model, representing the two models with the lowest AIC, there was a difference of 180 AIC. The result is illustrated in Fig. 4.6. Based on this outcome, the Extra-Trees model showed extreme differences and an enhanced chance of fit. AIC determines the model's convergence to actual data. However, over- or underfitting may persist. As such, further model assessment is required using other error metrics.

Out-of-sample data results were used to evaluate the models' performance. Since the models were trained using the training dataset, any attempt to duplicate the data will be highly accurate. As a result, the models are expected to have a high training accuracy score. However, a high level of training accuracy might lead to an unrepresentative model due to overfitting the data by matching inherent noise in the data. The proportion of accurate predictions generated by the models on the test data is used to assess the model performance on data that has not yet been observed. A high test accuracy is highly desired in the prediction of Vp and Vs to confirm model accuracy and robustness on new data since inaccurate measurements will have long-term defects in many operations. In this study, R^2 greater than 0.9 was regarded as an acceptable compromise between bias and variance. As can be seen in Fig. 4.7, the Extra-Trees model performed the best in terms of estimating Vp and Vs during testing.

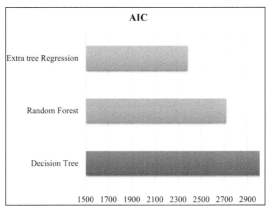

Fig. 4.6. AIC results comparing all the models on the test data.

Compressional and Shear Sonic Log Determination 77

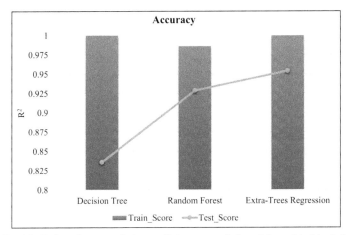

Fig. 4.7. Train and test correlation coefficient score of all models used in this study.

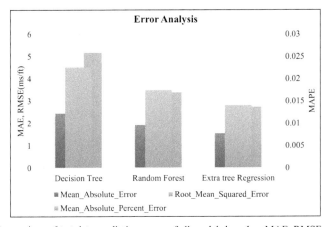

Fig. 4.8. Comparison of test data prediction errors of all models based on MAE, RMSE, and MAPE.

Any noteworthy improvement in accuracy attained by models significantly impacts decision-making. Although the models are from the same family, their theoretical and statistical foundations will give a wide range of model performances. Figure 4.8 displays the performance of the models on the test data to assess their precision, consistency, and robustness. The Extra-Trees model showed the most consistency, accuracy, and robustness when assessed using MAE, MAPE, and RMSE. The Extra-Trees thus surpassed the other models based on all the evaluation metrics.

Figure 4.9 shows the joint plot and regression combination used to visualise the predicted Vp and Vs test data and actual data and its error margins. The joint plot displays a univariate histogram of KDE curves and a bivariate cross-plot. Analysis of the regression plot and 95% confidence interval

78 Data Science and Machine Learning Applications in Subsurface Engineering

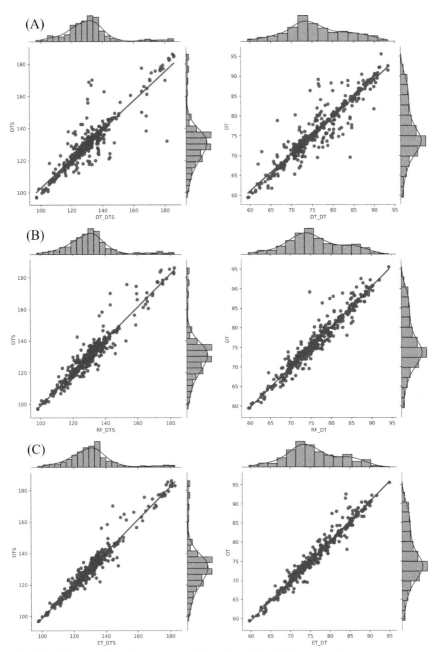

Fig. 4.9. A joint plot of Predicted vs Actual Vs (left) and Vp (right) on test data showing cross plot, data distribution, and confidence interval. (A) is Decision Tree, (B) is Random Forest and (C) is Extra-Trees.

illustrated in these figures indicates that the Extra-Trees model outperformed the other models. The Extra-Trees model showed a higher confidence interval and matched the range of values based on the histogram and KDE plot. Although all the models showed higher confidence in predicting Vp, the confidence level tends to deteriorate as high Vs values are predicted for the Decision Tree (above 140 μs/ft) and Random Forest (above 160 μs/ft) models.

The two key reasons that make Extra-Trees superior to Random Forests and Decision Trees are the combination of the separation of nodes by random selection of cut-points and the use of the entire training sample to reduce bias. Extra-Trees appropriately average the randomised predictions of Decision Trees to enhance precision while minimising computational complexity significantly. The Extra-Trees approach is built on the bias-variance concept. In addition to sharing similar advantages with the other models based on consistency for universal generalisation and approximation, Extra-Trees provide robustness in gross model errors since outliers have a small and local impact on its predictions. Extra-Trees may surpass Decision Trees in terms of computational efficiency by remaining impervious to irrelevant input features and when the input features exceed the random splits. The Decision Trees model exhibited a robust overfitting with 100% training accuracy but a relatively low out-of-sample accuracy compared to the Extra-Trees model. The Extra-Trees is then selected as the model with the most robust fit to the test data based on all the metrics.

6.2 Model Optimisation

The BO technique was employed to adjust its parameters to increase the Extra-Trees model's performance. Although the increase in accuracy was minimal, in subsurface engineering, any minimal increase in accuracy significantly impacts decision-making and can have a long-term effect on field development. Table 4.6 shows a comparison of all models against the BO-Extra-Trees (BO–ET) model. The predicted Vp and Vs were further evaluated based on formation type.

The results were further plotted in Fig. 4.10 and classified into rock types to evaluate the model's performance. The figure shows that the BO–ET model exhibited higher confidence in predicting Vs in both rock types. However, it was observed that most of the error is from the prediction of Vp and Vs in the sandstone interval. It is known that the sandstone interval is the hydrocarbon-bearing zone. As such, the effect of different fluids will influence the measurement of sonic velocities. Figure 4.11 shows a plot of the BO–ET predicted Vs and Vp compared to the empirical formulas Castagna and Greenberg-Castagna. The two empirical correlations only predict Vs, while the BO–ET predicts both Vs and Vp. The data were sorted based on

Table 4.6. Performance of all models compared to the Bayesian Optimised Extra-Trees.

R²	Decision Tree	|||||||
	Random Forest	
	Extra-Trees	||||||
	BO-Extra-Trees	|||||
MAE	Decision Tree	||||||||
	Random Forest	
	Extra-Trees	||||
	BO-Extra-Trees	|||
RMSE	Decision Tree	|||||||
	Random Forest	
	Extra-Trees	|||
	BO-Extra-Trees	||
MAPE	Decision Tree	||||
	Random Forest	
	Extra-Trees	||
	BO-Extra-Trees		

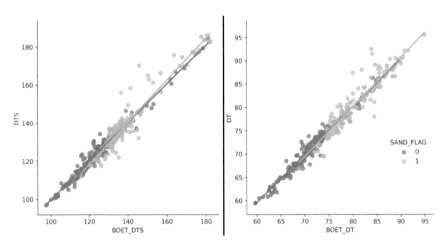

Fig. 4.10. Cross-plot of test and predicted data indicating confidence interval for BO–ET model based on formation type.

increasing saturation. By comparison, the BO–ET exhibited a better match done with the empirical correlations due to its data-driven approach. It was observed that at the zone of higher water saturation (mostly shales), the predicted Vp and Vs matched the actual better than the low water saturation zone (primarily sandstones). This observation confirms that different types

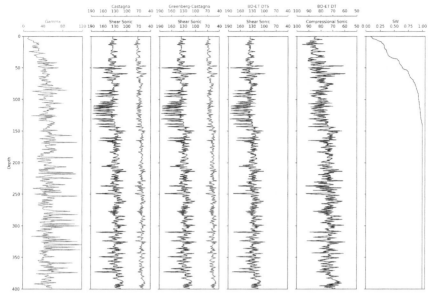

Fig. 4.11. Wireline log comparison of empirical and BO–ET predicted values against actual.

of fluids influence the measurements on sonic velocities, contributing to the model's error. The influence of the fluids can be adequately modelled to improve model confidence by retraining the model with additional data from the hydrocarbon-bearing interval.

The methodologies evaluated in this research demonstrated that a data-driven model is essential for Vs and Vp prediction. The shear and compressive sonic velocities are affected by various subsurface conditions. Wireline logs carry vital information about the subsurface. Hence, it is suitable to be used to measure Vs and Vp. The reason for most of the errors occurring in the sandstone interval is attributed to the effect of water saturation. There are significant variations in Vs and Vp as water saturation increases. When there is a high water saturation, grains tend to shear against one another. Elastic features can be seen in subsurface formations where the pore spaces are not completely saturated, allowing them to remain compressible. However, when more fluid percolates into a porous medium, the rock increases in rigidity, while elasticity reduces as the entire rock is occupied. The type of fluid or fluids coexisting in the pore spaces also influences rigidity and elasticity. The superior performance of the BO–ET model over the empirical correlations for Vs prediction is because the model is data-driven and considers both dry and saturated conditions and even the type of fluid saturation. There is an apparent interaction between fluids and the rock texture that influences Vs and Vp measurements.

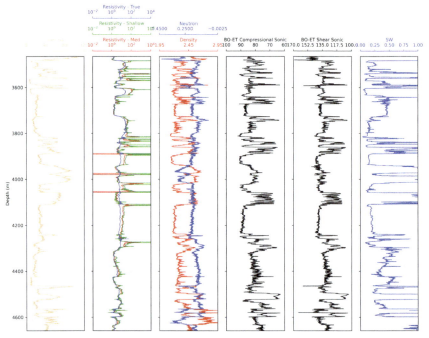

Fig. 4.12. Wireline log illustration of the BO–ET predicted Vp and Vs for the deployment Well C.

6.3 Model Deployment

After the successful evaluation of the BO–ET model, the model was retrained with all data from Well A and Well B. The retraining was necessary to ensure that the model learns from all the data sets. The BO–ET model was successfully deployed on Well C to predict Vs and Vp from the wireline logs. Figure 4.12 shows the wireline plot of the predicted Vs and Vp logs where the top reservoir is indicated at a depth of 3,570.6 m. The reservoir has three different formation water salinity in the three different formations. Based on observations made in the hydrocarbon-bearing zone, the predicted log within that interval was of high interest. The sensitivity towards fluid presence was adequately predicted for all three formations confirming the robustness of the model. This result demonstrates the appropriateness of the model for predicting Vs and Vp in similar formation types.

7. Conclusions

This study demonstrated the use of data-driven domain knowledge and an optimised Extra-Trees model for predicting shear and compressional sonic logs in wells that they are absent. This approach has been found to outperform empirical correlations that rely on the compressional sonic log to predict the shear sonic log in terms of accuracy. The assessment criterion indicated the models' likelihood of fitting held-out data. The models' accuracy, precision, and dependability were further evaluated based on the MAE, MAPE, RMSE, and R^2 on the held-out data. The findings are outlined as follows:

1. The input data were clearly justified to understand their influence on sonic velocity.
2. The Extra-Trees model outperformed the Random Forest and Decision Tree models in terms of precision, robustness, and probability of fit.
3. The BO algorithm made a marginal but meaningful improvement in the accuracy of the Extra-Trees model.
4. In the multi-output regression prediction, the predicted Vs in the sandstone interval exhibited the highest error on the test data.
5. The error in the sandstone interval is due to the influence of different formation fluids in the reservoir zone. This error was minimised by retraining the model with all the data from Wells A and B.
6. After deploying the model on Well C, the predicted Vs and Vp matched the influence of different fluid saturations in the three different formation types.

The proposed methodology offers the geothermal, fluid sequestration, and oil and gas industry a cheaper and time-efficient alternative to running the sonic logging tool in every drilled well. It also provides a way to measure Vs and Vp in older wells that were not previously measured.

Acknowledgement

The authors express their sincere appreciation to University Teknologi Petronas and the Centre for Subsurface Seismic Imaging for supporting this work.

Data Availability

The wireline log data used was obtained from Equinor Volve Field Datasets at https://www.equinor.com/en/how-and-why/digitalisation-in-our-dna/volve-field-data-village-download.html ([Dataset], 2018).

References

Ali, M., Jiang, R., Ma, H., Pan, H., Abbas, K., Ashraf, U. and Ullah, J. 2021. Machine learning: A novel approach of well logs similarity based on synchronization measures to predict shear sonic logs. *J. Pet. Sci. Eng.* 203: 108602. https://doi.org/10.1016/j.petrol.2021.108602.

Anemangely, M., Ramezanzadeh, A., Amiri, H. and Hoseinpour, S.A. 2019. Machine learning technique for the prediction of shear wave velocity using petrophysical logs. *J. Pet. Sci. Eng.* 174: 306–327. https://doi.org/10.1016/j.petrol.2018.11.032.

Asoodeh, M. and Bagheripour, P. 2012. Prediction of compressional, shear, and stoneley wave velocities from conventional well log data using a committee machine with intelligent systems. *Rock Mech. Rock Eng*. 45: 45–63. https://doi.org/10.1007/s00603-011-0181-2.

Asoodeh, M. and Bagheripour, P. 2014. ACE stimulated neural network for shear wave velocity determination from well logs. *J. Appl. Geophy*. 107: 102–107. https://doi.org/10.1016/j.jappgeo.2014.05.014.

Azizia, H., Siahkoohi, H.R., Evans, B., Farajkhah, N.K. and Kazemzadeh, E. 2017. A comparison between estimated shear wave velocity and elastic modulus by empirical equations and that of laboratory measurements at reservoir pressure condition. *Journal of Sustainable Energy Engineering* 5: 29–46. https://doi.org/10.7569/JSEE.2017.629502.

Bagheripour, P., Gholami, A., Asoodeh, M. and Vaezzadeh-Asadi, M. 2015. Support vector regression based determination of shear wave velocity. *J. Pet. Sci. Eng*. 125: 95–99. https://doi.org/10.1016/j.petrol.2014.11.025.

Breiman, L. 2001. Random forests. *Mach. Learn*. 45: 5–32. https://doi.org/10.1023/A:1010933404324.

Brocher, T.M. 2005. Empirical relations between elastic wavespeeds and density in the earth's crust. *Bulletin of the Seismological Society of America* 95: 2081–2092. https://doi.org/10.1785/0120050077.

Bukar, I., Adamu, M.B. and Hassan, U. 2019. A machine learning approach to shear sonic log prediction. *In*: Society of Petroleum Engineers–*SPE Nigeria Annual International Conference and Exhibition 2019*, NAIC 2019. Society of Petroleum Engineers, Lagos. https://doi.org/10.2118/198764-MS.

Carroll, R.D. 1969. The determination of the acoustic parameters of volcanic rocks from compressional velocity measurements. *International Journal of Rock Mechanics and Mining Sciences & Geomechanics Abstracts* 6: 557–579. https://doi.org/10.1016/0148-9062(69)90022-9.

Castagna, J.P., Batzle, M.L. and Eastwood, R.L. 1985. *Relationships between Compressional-wave and Shear-wave Velocities in Clastic Silicate Rocks*. pdf Vol. 50: 571–581.

Castagna, J.P., Batzle, M.L., Kan, T.K. and Backus, M.M. 1993. Rock physics—The link between rock properties and AVO response. Offset-dependent reflectivity—Theory and practice of AVO analysis. *SEG* 8: 135–171.

Chaikine, I.A. and Gates, I.D. 2020. A new machine learning procedure to generate highly accurate synthetic shear sonic logs in unconventional reservoirs. *In*: SPE Annual Technical Conference and Exhibition. SPE. https://doi.org/10.2118/201453-MS.

Close, D., Cho, D., Horn, F. and Edmundson, H. 2009. The sound of sonic: A historical perspective and introduction to acoustic logging. *Canadian Society of Exploration Geophysicists Recorder* 34: 34–43.

[Dataset]. 2018. Volve field data village download - data 2008-2016 - equinor.com [WWW Document]. URL https://www.equinor.com/en/what-we-do/digitalisation-in-our-dna/volve-field-data-village-download.html (accessed 5.27.21).

Dinov, I.D. 2018. *Data Science and Predictive Analytics*. Springer International Publishing, Cham. https://doi.org/10.1007/978-3-319-72347-1.

Duffaut, K. and Landrø, M. 2007. Vp/Vs ratio versus differential stress and rock consolidation—A comparison between rock models and time-lapse AVO data. *Geophysics* 72: C81–C94. https://doi.org/10.1190/1.2752175.

Dvorkin, J. and Mavko, G. 2014. V S predictors revisited. *The Leading Edge* 33: 288–296. https://doi.org/10.1190/tle33030288.1.

Eberli, G.P., Baechle, G.T., Anselmetti, F.S. and Incze, M.L. 2003. Factors controlling elastic properties in carbonate sediments and rocks. *The Leading Edge* 22: 654–660. https://doi.org/10.1190/1.1599691.

Elkatatny, S., Tariq, Z., Mahmoud, M., Mohamed, I. and Abdulraheem, A. 2018. Development of new mathematical model for compressional and shear sonic times from wireline log data using artificial intelligence neural networks (White Box). *Arab J. Sci. Eng.* 43: 6375–6389. https://doi.org/10.1007/s13369-018-3094-5.

Ernst, P., Wehenkel, D., Govindaraju, L. and Rao, R.S. 2006. Extremely randomized trees, *Artificial Neural Network in Hydrology*. Kluwer.

Folkestad, A. and Satur, N. 2008. Regressive and transgressive cycles in a rift-basin: Depositional model and sedimentary partitioning of the Middle Jurassic Hugin Formation, Southern Viking Graben, North Sea. *Sediment Geol.* 207: 1–21. https://doi.org/10.1016/j.sedgeo.2008.03.006.

Gamal, H., Alsaihati, A. and Elkatatny, S. 2022. Predicting the rock sonic logs while drilling by random forest and decision tree-based algorithms. *J. Energy Resour. Technol.* 144. https://doi.org/10.1115/1.4051670.

Gassmann, F. 1951. Elastic waves through a packing of spheres. *Geophysics* 16: 673–685. https://doi.org/10.1190/1.1437718.

Geurts, P., Ernst, D. and Wehenkel, L. 2006. Extremely randomized trees. *Machine Learning* 63(1): 3–42. https://doi.org/10.1007/s10994-006-6226-1.

Gordon, A.D., Breiman, L., Friedman, J.H., Olshen, R.A. and Stone, C.J. 1984. Classification and regression trees. *Biometrics* 40: 874. https://doi.org/10.2307/2530946.

Greenberg, M.L. and Castagna, J.P. 1992. Shear-wave velocity estimation in porous rocks: theoretical formulation, preliminary verification and applications. *Geophys. Prospect.* 40: 195–209. https://doi.org/10.1111/j.1365-2478.1992.tb00371.x.

Hamada, G. and Joseph, V. 2020. Developed correlations between sound wave velocity and porosity, permeability and mechanical properties of sandstone core samples. *Petroleum Research* 5: 326–338. https://doi.org/10.1016/j.ptlrs.2020.07.001.

Hatampour, A. and Ghiasi-Freez, J. 2013. A fuzzy logic model for predicting dipole shear sonic imager parameters from conventional well logs. *Pet. Sci. Technol.* 31: 2557–2568. https://doi.org/10.1080/10916466.2011.603005.

LeCompte, B., Majekodunmi, T., Staines, M., Taylor, G., Zhang, B., Evans, R. and Chang, N. 2021. Machine learning prediction of formation evaluation logs in the Gulf of Mexico. *In*: *Offshore Technology Conference*. OTC. https://doi.org/10.4043/31093-MS.

Nourafkan, A. and Kadkhodaie-Ilkhchi, A. 2015. Shear wave velocity estimation from conventional well log data by using a hybrid ant colony–fuzzy inference system: A case study from Cheshmeh–Khosh oilfield. *J. Pet. Sci, Eng.* 127: 459–468. https://doi.org/10.1016/j.petrol.2015.02.001.

Onalo, D., Adedigba, S., Khan, F., James, L.A. and Butt, S. 2018. Data driven model for sonic well log prediction. *J. Pet. Sci. Eng.* 170: 1022–1037. https://doi.org/10.1016/j.petrol.2018.06.072.

Otchere, D.A., Arbi Ganat, T.O., Gholami, R. and Lawal, M. 2021a. A novel custom ensemble learning model for an improved reservoir permeability and water saturation prediction. *J. Nat. Gas Sci. Eng.* 91: 103962. https://doi.org/10.1016/j.jngse.2021.103962.

Otchere, D.A., Arbi Ganat, T.O., Gholami, R. and Ridha, S. 2021b. Application of supervised machine learning paradigms in the prediction of petroleum reservoir properties: Comparative analysis of ANN and SVM models. *J. Pet. Sci. Eng.* 200: 108–182. https://doi.org/10.1016/j.petrol.2020.108182.

Otchere, D.A., Ganat, T.O.A., Ojero, J.O., Taki, M.Y. and Tackie-Otoo, B.N. 2021c. Application of gradient boosting regression model for the evaluation of feature selection techniques in improving reservoir characterisation predictions. *J. Pet. Sci. Eng.* 109244. https://doi.org/10.1016/J.PETROL.2021.109244.

Otchere, D.A., Abdalla Ayoub Mohammed, M., Ganat, T.O.A., Gholami, R. and Aljunid Merican, Z.M. 2022a. A novel empirical and deep ensemble super learning approach in predicting reservoir wettability via well logs. *Applied Sciences* 12: 2942. https://doi.org/10.3390/app12062942.

Otchere, D.A., Ganat, T.O.A., Nta, V., Brantson, E.T. and Sharma, T. 2022b. Data analytics and Bayesian Optimised Extreme Gradient Boosting approach to estimate cut-offs from wireline logs for net reservoir and pay classification. *Appl. Soft. Comput.* 120: 108680. https://doi.org/10.1016/j.asoc.2022.108680.

Phadke, S., Bhardwaj, D. and Yerneni, S. 2000. Marine Synthetic Seismograms Using Elastic Wave Equation. *2000 SEG Annual Meeting*, Calgary, Alberta.

Pickett, G.R. 1963. Acoustic character logs and their applications in formation evaluation. *Journal of Petroleum Technology* 15: 659–667. https://doi.org/10.2118/452-PA.

Rasouli, V., Pallikathekathil, Z.J. and Mawuli, E. 2011. The influence of perturbed stresses near faults on drilling strategy: A case study in Blacktip field, North Australia. *J. Pet. Sci. Eng.* 76: 37–50. https://doi.org/10.1016/j.petrol.2010.12.003.

Rojas, E. 2008. Vp-Vs ratio sensitivity to pressure, fluid, and lithology changes in tight gas sandstones. *First Break* 26. https://doi.org/10.3997/1365-2397.26.1117.27907.

Safaei-Farouji, M., Hasannezhad, M., Rahimzadeh Kivi, I. and Hemmati-Sarapardeh, A. 2022. An advanced computational intelligent framework to predict shear sonic velocity with application to mechanical rock classification. *Sci. Rep.* 12: 5579. https://doi.org/10.1038/s41598-022-08864-z.

Simm, R. and Bacon, M. 2014. *Seismic Amplitude*. Cambridge University Press. https://doi.org/10.1017/CBO9780511984501.

Sneider, J.S., Clarens, P. de and Vail, P.R. 1995. Sequence stratigraphy of the middle to upper jurassic, viking graben, North sea. *Norwegian Petroleum Society Special Publications* 5: 167–197. https://doi.org/10.1016/S0928-8937(06)80068-8.

Suleymanov, V., Gamal, H., Glatz, G., Elkatatny, S. and Abdulraheem, A. 2021. Real-time prediction for sonic slowness logs from surface drilling data using machine learning techniques. *In: SPE Annual Caspian Technical Conference*. SPE. https://doi.org/10.2118/207000-MS.

Tarafder, S., Badruddin, N., Yahya, N. and Egambaram, A. 2021. EEG-based drowsiness detection from ocular indices using ensemble classification. pp. 21–24. *In: 2021 IEEE 3rd Eurasia Conference on Biomedical Engineering, Healthcare and Sustainability (ECBIOS)*. IEEE, https://doi.org/10.1109/ECBIOS51820.2021.9510848.

Tarafder, S., Badruddin, N., Yahya, N. and Nasution, A.H. 2022. Drowsiness detection using ocular indices from EEG signal. *Sensors* 22: 4764. https://doi.org/10.3390/s22134764.

Tosaya, C. and Nur, A. 1982. Effects of diagenesis and clays on compressional velocities in rocks. *Geophys. Res. Lett.* 9: 5–8. https://doi.org/10.1029/GL009i001p00005.

Waluyo, W., Uren, N.F. and McDonald, J.A. 1995. Poisson's ratio in transversely isotropic media and its effects on amplitude response: An investigation through physical modelling experiments. pp. 585–588. *In: SEG Technical Program Expanded Abstracts 1995*. Society of Exploration Geophysicists. https://doi.org/10.1190/1.1887420.

Zoveidavianpoor, M., Samsuri, A. and Shadizadeh, S.R. 2013. Adaptive neuro fuzzy inference system for compressional wave velocity prediction in a carbonate reservoir. *J. Appl. Geophy*. 89: 96–107. https://doi.org/10.1016/j.jappgeo.2012.11.010.

CHAPTER 5

Data-Driven Virtual Flow Metering Systems

*Ramez Abdalla** and *Philip Jaeger*

1. Introduction

Proper estimation of multiphase flowrates in oil and gas production systems is an essential tool of monitoring and optimising the production systems. Hence, one of the routine tests of wells is production testing. It is usually conducted as a schedulable test to monitor liquid rates, water cuts, and gas oil ratio (GOR). The production testing is easily conducted using the test separator to compare the actual production rate with the theoretical one. It is the most common form of production and reservoir surveillance. However, this technique has its limitations. The main limitation of this test is the insufficient resolution or repeatability to identify trends in liquid and water-cut rates over short periods of time. Another potential problem could be the duration issue. It is often the case in low-flow rate and deep wells, which require several time-consuming whole or complete liquid holdup periods. Later, an alternative solution to the production testing has been developed. This solution is called multiphase physical flow meters (MPFMs). This technology depends on the idea of indirectly estimating multiphase flowrates without separating the phases. This is done by tracking supplementary measurements of fluid phase properties such as velocity and phase fractions inside the device. These meters are usually installed at the wellhead, so that the multiphase flowrates of a particular well can be tracked in real-time. One disadvantage of MPFMs is that they have higher capital costs (CAPEX) and operating costs (OPEX), and they require frequent production calibration.

Clausthal University of Technology, Clausthal, Germany.
* Corresponding author: ramez.maher.aziz.zaky.abdalla@tu-clausthal.de

Subsequently, the technology of virtual flow metering (VFM) arises as an attractive technology in the oil and gas industry. It depends on either analytical or data-driven models for real-time calculations of phases' production. In this chapter, data driven virtual flow metering solutions are presented. At the beginning, VFM classification is introduced based on modelling paradigms. They are classified as first principle and data-driven VFMs. Then, an overview of the applications of data-driven modelling in VFM systems is presented. Through discussing these applications, the model's features used in the works, the predicted variables, the input data for the training, and the respective paper are emphasised.

In addition, the data driven virtual flow meter components and methodology to develop this type of flow meter is discussed. The methodology follows the cross-industry standard process for a data mining framework. This is a framework for project planning that many industrial projects have used. Its main components are data understanding and preparation, modelling, and evaluation. Subsequently, an implementation of this methodology is presented to estimate flow rate and water-cut prediction from the electrical submersible pump's sensor data. Finally, we introduce a field experience with a data-driven VFM system. In this section, the real operational experience reported in the literature using the data-driven models are discussed.

2. VFM Key Characteristics

The near wellbore area, wells, pipelines, and production chokes must be simulated to be able to formulate systems. To come up with precise estimates of flowrates, models are combined with observations like inlet and outlet pressures and temperatures, choke openings, etc. Depending on the measurement data available, the production system can either be represented as a whole from the reservoir to the processing plant, or it can be divided into sub models. Figure 5.1 shows a schematic graph of well sub models and relevant data. Also, optimisation algorithms may be used to modify flowrates and other tuning parameters. This is mainly to stabilise estimation models by reducing the discrepancy between model predictions and actual measurements.

The concept behind these systems makes them have general characteristics in common whatever the system sub modules are. One of these characteristics is that they are dependent on instrument and system sensitivity to changes in flow and phase fractions. They are also dependent on tuning/calibration to extend to a "calibrated range" and adapt to a changing operating condition. They can be classified into pure data-driven or thermo-hydraulic modelling VFMs, or hybrids. All of them, either data-driven or mechanistic, should take into account pressure and temperature drop in the well, and pressure drop over choke and phase fractions of the flowing stream.

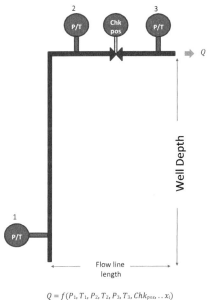

$$Q = f(P_1, T_1, P_2, T_2, P_3, T_3, Chk_{pos}, \ldots x_i)$$

Fig. 5.1. Schematic graph of well sub models.

Many conservation equations in the mechanistic VFMs take a dynamic form. However, because of the steady state or quasi-steady state nature of the optimisation problem formulation, an optimisation solver can only discover a solution for a single point in time or can use the solution from the previous step as a first estimate for the prediction of the current time step. Sometimes, due to the nature of the problem, the conservation equations adopt steady state forms or do not take time as a variable parameter. Although it is feasible to frame the VFM as an optimisation problem dynamically, such a strategy is not applied in VFM literature. The primary reason for this could be the cost of complex computations of the dynamic optimisation for VFM systems. Also, that would not make it suitable for real-time applications.

In addition to dynamic optimisation, Kalman filter approaches and other state estimation methods may be employed to develop a dynamic VFM. The major difficulty of this method could be the high level of knowledge required for setup and use of this application, as well as the difficulty of tuning the actual field data in a reliable way.

On the other hand, the data-driven VFM technique is based on gathering field data and mathematically adapting it to the production system's physical parameters, such as well bore and choke geometry, flow-line wall thickness, etc., without providing an explicit description of those parameters. Over the past few years, this strategy has gained a lot of popularity not just for oil and gas applications but also for many other applications. The data-driven model can conduct quick and precise real-time metering if the model has been properly trained and the exposed conditions fall within the training range. This method can build models more affordably than mechanistic models since it does not require much in-depth production engineering domain expertise. In the following section, we are presenting various data-driven VFM applications on different oil wells systems.

3. Data Driven VFM Main Application Areas

Virtual sensors are convenient replacements of physical sensors that use available data during well-known conditions in instrumented wells to predict other measurements (Vinogradov and Vorobev, 2020). For instance, wellhead pressure (WHP) and wellhead temperature (WHT) are related to flow line pressure (FLP) for a specific choke diameter; therefore, it is convenient to establish a machine learning model among these four variables so that it can serve as a replacement whenever needed. Since most of the oil and gas wells exhibit a non-stationary process, where the boundary conditions may change along with the life of the field, it is probably safe to recognise that the virtual sensor model may not be valid throughout the life of the well, but for a specific period of time. Finally, a data-driven VFM would be very beneficial when there is sufficient measured data, including frequent well tests (e.g., 8–12 per year), permanent wellhead and flow line sensors (pressure, temperature). In the upcoming subsections, we are introducing different attempts of VFM for various production systems.

3.1 Virtual Sensing in ESP Wells

Electric submersible pumps (ESPs) are currently widely employed on many artificially lifted wells with high water cut and offshore oil wells due to their simple structure and high efficiency (Takacs, 2018). Among all the artificial lift systems, ESP is preferred because it can produce high volumes in higher temperatures and reach deeper depths. The development of sensors and data acquisition systems make it possible for ESP systems to continuously record the intake pressure and temperature, pump head, discharge pressure, and temperature, motor temperature and current, leakage current, vibration, etc. These data would be recorded at regular intervals and transmitted to surface remote terminal units (RTUs) (Carpenter, 2019). Figure 5.2 shows a schematic graph of sub-modules and sensors deployed on the electrical submersible pump well system.

Based on the advancement in communication, well technology, and field equipment, a lot of approaches are made for the estimation of flow rates in real time. In the following, a summary of the applications of soft sensing both mechanistic and surrogate modelling is given.

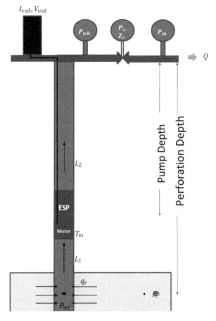

Fig. 5.2. A schematic diagram of an ESP well.

3.2 Virtual Sensing for SRP Wells

Beam pumping, or the sucker-rod lift method, is the oldest and most widely used type of artificial lift. A sucker-rod pumping system is made up of several components, some of which operate on the surface and other parts operate underground, down in the well. The surface-pumping unit, which drives the underground pump, consists of a prime mover (usually an electric motor) and, normally, a beam fixed to a pivotal post which is called a Sampson post and the beam is called a walking beam.

Several sensors can provide measurements of sucker-rod pump operations. The main measurements are loads on the pump which form what are called dynamomter cards which are diagnostic cards that plot the load on the top rod (polished rod) in relation to the polished rod position as the pumping unit moves through each stroke cycle. The plot of the polished rod Load vs Position is known as the Surface Card. Subsequently, a wave equation solution is used to derive the downhole card from the surface card. The downhole card is a plot of the Load vs Position on the pump's plunger. Also, on the surface, continuous measurements for wellhead pressures, temperatures and power measurements of the motor are reported. In addition, the normal frequent tests data for any production system are provided. Those are fluid level depth using acoustic transducer and production of multi phases using a separator test or

Table 5.1. Applications of the VFM on ESP wells.

Author	Model Summary
Camilleri et al. (2011, 2015, 2016, 2016b, 2017)	In these papers, ESP models with different modifications are discussed, as well as field case studies where the ESP's first principle models serve as virtual flow meters. A hybrid method has been used to measure the flow rate without needing a test separator or multi-phase flowmeter. It is stated that the pump absorbs an amount of power equating that is generated by the motor. The drop of the pressure in the tubing provides measurements of the average density of the fluid, which is then converted to a water cut. Obtaining a reference for validating the calculation, a comparison has been carried out between the calculated and the measured rates on a shale oil well equipped with an ESP (Camilleri, 2011, 2015, 2016a, 2016b, 2017).
Haouche et al. (2012a, 2012b)	The VFM model is a combination of three main units: Reservoir unit, Electrical Submersible Pump unit, and Production Tubing unit. A density correction factor is used to take into account the effects of gas on the operational performance of the submersible pump (Adrien, 2012; Haouche, 2012).
Binder et al. (2015)	A Moving Horizon Estimator is applied for flowrate estimation in a well with an ESP. As input, the bottomhole, the downhole, and the pump pressure sensors are considered together with pump parameters. The method showed an accurate performance and was suggested to be used for industrial applications (Binder et al., 2015).
Zhu et al. (2016)	In this research, singular spectrum analysis (SSA) is used on a raw production dataset without any pre-processing or transformation of the original series. They investigated the decomposition of the original series into a summation of the principal independent and interpretable components such as slowly varying trends, cycling components, and random noise (Zhu et al., 2016).
Krikunov et al. (2019)	A hybrid physical-machine learning prediction model was developed. It utilised a range of motor frequencies. A numeric optimisation model was created to suggest multi-well operating modes (Krikunov et al., 2019).
Zhu et al. (2020, 2021)	In these studies, a mechanistic model was developed to predict the pump-boosting pressure. The objectives were to forecast oil-water emulsion rheology and how it will affect ESP boosting pressure and describe the pump leaking impact under the conditions of a gas-liquid flow (Zhu et al., 2020, 2021).
Sabaa et al. (2022)	This study aims to develop artificial neural network models to predict flow rates of ESP artificially lifted wells. Each data set included measurements for wellhead parameters, fluid properties, ESP downhole sensor measurements, and variable speed drive (VSD) sensor parameters. The models consisted of four separate neural networks to predict oil, water, gas, and liquid flow rates (Sabaa et al., 2022).

production test. Figure 5.3 shows the sub-modules of sucker rod pumped well with relevant tests, dynamometer, and fluid level test example.

The applications of virtual sensing on the sucker rod pump systems known so far are limited. This may be due to its limited ability in producing high fluid rates. Some attempts have been made in this regard. Those applications had various objectives. The first objective is predicting multi phase flow rates or the dynamic fluid level in the annulus using dynamomter cards and well head pressure and temperature as inputs. The second objective is inferring the dynamometer cards using electrical power data.

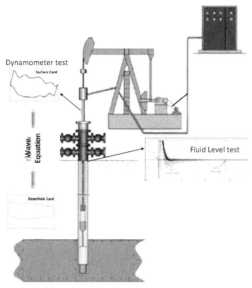

Fig. 5.3. A schematic graph of an SRP well.

3.2.1 Virtual Flow Meter on Rod Pumping Systems

The challenge of predicting oil, gas and water flow is a function that describes the multi phase flow rates. As a solution, data-driven algorithms are used to find a relation between the pump operational parameters and the produced oil, gas, and water. Peng et al. (2019) have used the deep autoencoders to derive features from dynamometer cards to generate a predictive model for production. Combining this model with pump and production data leads to abstract features that show good accordance with the history data (Peng et al., 2019).

Soft sensing, which replaces the traditional detection method to model the dynamic liquid level of the sucker-rod pumping system (Haitao et al., 2014), proposes a method to calculate the dynamic fluid level. It takes the submerged pressure as a common solution node to analyse both the plunger load variation which is contributed by the pump dynamometer card and the pressure distribution in the annulus. Li et al. (2013) presented the simulated annealing based on a Gaussian regression modeling.

3.2.2 Virtual Sensing of the Dynamometer Card

Through the years, many researchers have been studying the relation between electrical parameters and surface cards (Zhang and Tang, 2008). Some theory formulas also can be built to calculate the card from the electrical parameters. Nevertheless, some parameters in the formulas cannot be quantified or measured and some assumptions of the values made the card calculation inaccurate and unstable. Thus, a machine learning model for dynamometer card calculation in the rod pumping lift process is used to formulate the complicated process. Deep neural networks are used to find good weight combinations in these examples that eventually allow the model to come up with rules from the input data (electrical parameter) to the target data (dynamometer cards).

Such a study includes extracting power features and constitutes an eigenvector in chronological sequence for one period. Afterwards, dynamometer card data and shape curve image are extracted according to coordinates and load data. Subsequently, power features and dynamometer diagram features are normalized by rows and then mapped between 0 and 1. Finally, a sequence-to-sequence algorithm to infer dynamometer cards from the power curve features (Peng et al., 2019; Zhu et al., 2021).

3.3 Virtual Sensing for Gas Lifted Wells

Gas lift (GL) is a method of artificial lift that uses an external source of high-pressure gas for supplementing the formation of gas to lift the well fluids. The principle of the GL is that gas injected into the tubing reduces the density of the fluids in the tubing and the bubbles have a 'scrubbing' effect on the liquids. Both factors act to lower the flowing bottomhole pressure (BHP) at the bottom of the tubing.

One use of the case for VFM modelling is to compute uncorrelated estimates of the GL rate by using the GL flow control valve performance model.

Figure 5.4 shows the control volumes of the gas lifted well. In this system, the inputs are manifold pressure, valve and casing pressure, while the output is GL rates.

Applications of this research by Ziegel et al. (2014) discuss the design of a data-driven model that would support and optimize production operations during normal or gas coning. The data-driven model had a limited set of inputs such as production choke, GL choke, resulting GL and reservoir pressure estimation. Therefore, several predictive models will be built to create a final data-driven model (Ziegel et al., 2014).

Iman et al. (2020) developed a data-driven approach to find an optimal operating envelope for GL wells. The process involves building a multilayer perceptron neural network model for generating instantaneous predictions of multiphase flow rates and other quantities of interest, such as GOR, WCT, using real-time sensor data at the surface, historical performance and sporadic test data. The models were developed for generating short-term (30 days) forecast of cumulative oil, water, gas and liquid production, multiphase flow

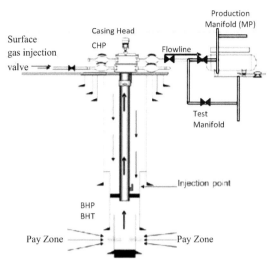

Fig. 5.4. A schematic graph of a GL well.

rates, WCT, GOR, and reservoir pressure. Using time-series forecasting models, a sensitivity analysis was performed to generate short-term well response for a selected number of combinations of choke settings and gas injection rates (Al Sebaiti et al., 2020).

Khan et al. (2020) used AI techniques to derive a robust correlation for forecasting production rates in gas-lift assisted wells. The AI techniques used in this research included artificial neuro-fuzzy inference systems, artificial neural network (ANN), functional networks, and support vector machines. They collected test data from several GL wells and used ANN to develop an equation to forecast oil flow rate.

Initially, they applied wide data analytics and then input data to the models that were compared to each other and to other empirical models. They could predict oil rates with an accuracy exceeding 98% (Khan et al., 2020).

3.4 Virtual Sensing for Gas Wells and Plunger Lifted Wells

Conventional plunger lifting is a transient process that consists of cyclic openings and closings of a gas well. Because of their complex behaviour, using traditional physics-based models to simulate the coupled behaviour of reservoir and wellbore performance is computationally rigorous and challenging. Therefore, machine learning methodology would help in formulating the plunger lifted well system including plunger arrival time, tubing and casing pressure, and instantaneous gas flow rate. The sequence of stages in one cycle. The motor valve has only two states, namely closed or open. The length of each state depends on the pre-programmed triggers,

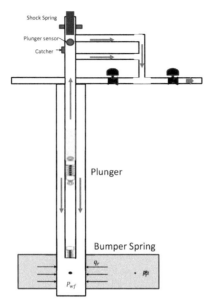

Fig. 5.5. A schematic graph of a plunger lifted well.

which are also called controller set-points. The main triggers to close a well include a fixed timer for the after-flow stage, the difference between the gas and the calculated critical flow rates, as well as various relations between casing, tubing, and line pressures. The main triggers to open a well include a fixed timer for total "off time", calculated average plunger rise velocity, and different relations between the pressure measurements. Figure 5.5 shows the control volumes of a plunger lifted well.

As aforementioned, virtual flow metering includes physics-based and data-driven methods. When it comes to the plunger lift application, since the process is extremely transient, the application of physics-based methods is extremely complex and unviable (Akhiiartdinov et al., 2020). Regarding data-driven models, Andrianov (2018) (GarcAa et al., 2010) and Shoeibi Omrani et al. (2018) (Loh and Omrani, 2018) demonstrated the application of ANNs to simulate the transient behaviour of severe slugging and liquid-loaded gas wells. On the other hand, Akhiiartdinov et al. (2020) have attempted to model a VFM on plunger lifted wells. In this study, the objective was to optimize the "on" and "off" periods of the control valve, which serve as parameters for building the response surface.

3.5 Miscellaneous Applications for Identifying Flow Regimes

As aforementioned, there are common characteristics of VFMs. One of the main characteristics is estimation of the fraction of each phase or, in other

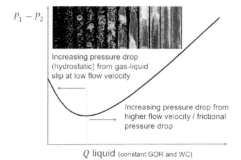

Fig. 5.6. Multiple-phase flow regimes.

words, identifying the flow regime. Therefore, various applications arise in this area. Their objective is to construct a classification model to predict various flow regimes based on flow measurements. Their work focuses on the identification of multiple-phase flow regimes by implementing deep learning algorithms in addition to commonly used machine learning algorithms (Alhashem, 2020; Arteaga-Arteaga et al., 2021; Manikonda et al., 2018; Mask et al., 2019). Figure 5.6 shows the relation between the pressure drop versus flow rate with different flow regimes.

4. Methodology of Building Data-driven VFMs

4.1 Data Collection and Preprocessing

Data collection and preprocessing are two main steps in constructing any data-driven model. Usually, the data is collected from several locations in the field or historical data from other relevant wells or fields. After collecting the data set that will be used to train the data-driven model, this data set usually requires processing since it may be noisy, corrupted, or has missing values, deviations, and irrelevant data. Therefore, the preprocessing phase of the data set is essential to clean and authenticate it before training the model.

In the preprocessing step, the data set can undergo some modifications and more information can be gathered from it. Applying feature engineering techniques in the preprocessing step will support constructing the data-driven model to locate the complex interactions between the original data and the output or eliminate unnecessary features which increase the computational costs. One common method of feature engineering is dimensional reduction using principal component analysis (PCA), feature selection methods, and linear or nonlinear combinations of raw features.

In general, feature engineering may help the data-driven algorithm to find complex relationships between the original data and the output variable or remove redundant features which lead to lower computational cost during training and prediction steps.

In most of the cases, subject matter experts (SME) construct informative features for further algorithm training. Creating good features using the input data which can describe the multiphase flow transport process may help to obtain better predictions.

4.2 Model Development

Model Development is the phase in which the implemented algorithm constructs a map of the input and output features. This mapping process is also called the training or learning phase of the model. In this phase, the model will modify the available parameters to precisely predict the final output. The technique the parameters are modified differs depending on the algorithm implemented in building the model. The next step in the model development is verifying and testing the model on different data sets to see how the model will perform on untrained data.

Additionally, the validation step is important to adjust the hyper-parameters. Hyper parameters are selected before training to decrease noise and deviations that reduce the prediction precision of the model. K-fold cross-validation is a new technique in which the data set is divided into training and testing data then the model hyper parameters are assigned. Then the model will be trained on K-folds and the average error is calculated for the model. The K-fold will be repeated with several hyper parameters to find the best ones with the lowest average error.

The final phase is to confirm the performance of the validated model on a separate data set. In the final stage, there are two possible outcomes: first, the errors on the training and testing sets are large which is called underfitting, and second, the errors on the training are small but large on the testing set which is called overfitting. Thus, validating and testing the model to assign the best hyper parameters to figure out the finest outcome is the optimal goal. A typical choice for the number of folds to use would be k = 10. Figure 5.7 shows a 10-fold cross-validation.

Fig. 5.7. 10-fold cross-validation.

A common algorithm used for building VFM models is the ANN which could create a function originating from the inputs that connect them to the final output. Generally, an ANN consists of an input layer to read the input features which usually are pressure, temperature, choke opening, or other production system parameters in the case of a VFM.

The second component of the ANN is the hidden layer where non linear functions are created to build the connection between the inputs and the prediction and the final component is the output layer where the results achieved in the hidden layers go through an activation function to reach the final prognosis. The ANN showed significant results when applied to a steady state flow unlike the case of transient movement.

Al-Qutami et al. (2017) showed in three publications the application of ANN in constructing a VFM model (Al-Qutami et al., 2017, 2017, 2017). Their work provides a significant sighting to improve VFM models. The latest methodology applied in their research was an ANN model was trained using the Levenberg-Marquardt optimisation algorithm and K-fold cross-validation to specify the number of neurons. This work was able to conclude that the gas flow rate is most sensitive to choke opening and that the bottom hole pressure is the most critical feature for the flow rate prediction.

Loh and Omrani (2018) implemented feed-forward neural networks on simulated and field data to predict oil and gas flow rates. The results showed that the NN model was effective at steady state flow while the model produced transient flow imprecise predictions. Furthermore, they conducted a sensitivity study on the input features that showed that the NN model was able to perform well even in case of noisy inputs. Finally, they introduced a new back-allocation technique to predict the flow rate using separator measurements. Alajmi et al. (2015) introduced an NN model to predict the oil flow rate through the choke and they added an empirical correlation for critical choke flow to the input features. The study demonstrated a noticeable enhancement in the NN model performance compared to the mechanistic models.

To conclude in this section, we have reviewed the main building blocks of creating VFMs. It started with data gathering, preprocessing, and feature engineering. We have also presented the validation techniques. Finally, we presented some of NN applications to VFM as an example for the algorithms used for data-driven VFMs.

5. Field Experience with a Data-driven VFM System

In the pursuit of VFMs, several industrial and research works were published. In this section, we focus on highlighting the industrial applications that show a bench marking or a commercial tool. Among the available industrial works available Denny et al. (2013) reported the importance of integrating VFM in an electrical submersible pump model. GarcAa et al. (2010) reported another solution for monitoring production and injection rates using an NN model instead of fiscal meters on individual wells. NN is trained by sensors, well-tests, and simulation data. Their work results showed a 4% error in the overall wells flow-rates' predictions compared to the fiscal meters.

Moreover, Olivares et al. (2012) describe the implementation of an NN model and nodal analysis joint with sporadic and real-time data to determine and observe daily flow rates (Olivares et al., 2012). The new applied methodology proved to be more efficient than the back allocation system. Al-Jasmi et al. (2013) report the application of a flow-rate prediction NN model on a mature carbonate reservoir in the Middle East (Cramer and Goh, 2009; Cramer et al., 2011; Goh et al., 2008; Law et al., 2018; Poulisse et al., 2006). Those papers describe Shell's Data-Driven model FieldWare Production Universe (FW PU), which was implemented in several case studies to estimate flow rates and optimise the production operation. In the six study cases, FW PU was able to enhance productivity and reduce the amount of gas used in the GL which significantly affected the cost.

Bello et al. (2014) also used LR with feature extraction PCA which showed significant results for oil and gas flow-rate predictions. Grimstad et al. (2016) used pressure drop, choke, and inflow performance models from prosper to train a B-spline surrogate model. The results showed reasonable performance when compared to OLGA.

References

Adrien, M.H., Younes, T., Deffous, J-F., Couput, A.J-P., Caulier, R. and Vrielynck, B. 2012. Smart metering: An online application of data validation and reconciliation approach. *Paper presented at SPE Intelligent Energy International Utrecht, The Netherlands, March.* pp. SPE–149908–MS.

Akhiiartdinov, A., Pereyra, E., Sarica, C. and Jose Severino, J. (eds.). 2020. Data analytics application for conventional plunger lift modelling and optimisation, volume Day 1 Tue., 10 November. *SPE Artificial Lift Conference and Exhibition – Americas*.

AlAjmi, Mohammed D., Abdulraheem, A., Mishkhes, A.T. and Al-Shammari, M.J. (eds.). 2015. Profiling downhole casing integrity using artificial intelligence, volume Day 1 Tue., 03 March *SPE Digital Energy Conference and Exhibition*.

Alhashem, M. (ed.). 2020. Machine learning classification model for multiphase flow regimes in horizontal pipes, volume Day 2 Tue., 14 January *IPTC International Petroleum Technology Conference*.

Al-Jasmi, A., Goel, H.K,. Nasr, H., Querales, M., Rebeschini, J., Villamizar, M.A., Carvajal, G.A., Knabe, S., Rivas, F. and Saputelli, L. 2013. Short-term production prediction in real time using intelligent techniques. *Paper presented at EAGE Annual Conference and Exhibition at London, UK. In*: *All Days*, SPE. pp. SPE–164813–MS.

Al-Qutami, T.A., Ibrahim, R., Ismail, I. and Ishak, M.A. 2017a. Development of soft sensor to estimate multiphase flow rates using neural networks and early stopping. *International Journal of Smart Sensing and Intelligent Systems* 10(1): 1–24.

Al-Qutami, T.A., Ibrahim, R. and Ismail, I. 2017b. Hybrid neural network and regression tree ensemble pruned by simulated annealing for virtual flow metering application. pp. 304–309. In: *2017 IEEE International Conference on Signal and Image Processing Applications (ICSIPA)*.

AL-Qutami, T.A., Ibrahim, R., Ismail, I. and Ishak, M.A. 2017c. Radial basis function network to predict gas flow rate in multiphase flow. pp. 141–146. In: *Proceedings of the 9th International Conference on Machine Learning and Computing*, ACM.

Al Sebaiti et al. 2020. Robust data-driven well performance optimisation assisted by machine learning techniques for natural flowing and gas-lift wells in Abu Dhabi, volume Day 4 Thu., 29 October, of *SPE Annual Technical Conference and Exhibition*. D041S046R002.

Arteaga-Arteaga, H.D., Mora-Rubio, A. Florez, F., Murcia-Orjuela, N., Diaz-Ortega, C.E., OrozcoArias, S., delaPava, M., Bravo-OrtAz, M.A., Robinson, M., Pablo Guillen-Rondon, P. and Tabares-Soto, R. 2021. Machine learning applications to predict two-phase flow patterns. *Peer J. Comp. Sc.* 7: e798.

Bello, O., Ade-Jacob, S. and Kun Yuan, K. (eds.). 2014. Development of hybrid intelligent system for virtual flow metering in production wells, volume *All Days*. *SPE Intelligent Energy International Conference and Exhibition*.

Binder, Benjamin J.T., Pavlov, A., Tor, A. and Johansen, T.A. 2015. Estimation of flow rate and viscosity in a well with an electric submersible pump using moving horizon estimation. This work is funded by the Research Council of Norway and Statoil through the petromaks project no. 215684: Enabling high-performance safety-critical offshore and subsea automatic control systems using embedded optimization (emopt). *IFACPapersOnLine* 48(6): 140–146. *2nd IFAC Workshop on Automatic Control in Offshore Oil and Gas Production OOGP 2015*.

Camilleri, L. and Zhou, W. (eds.). 2011. Obtaining real-time flow rate, water cut, and reservoir diagnostics from ESP gauge data, volume *All Days*. SPE *Offshore Europe Conference and Exhibition, 2011*.

Camilleri, L., El Gindy, M., Rusakov, A. and Adoghe, S. (eds.). 2015. Converting ESP real-time data to flow rate and reservoir information for a remote oil well, volume Day 2 Wed., 16 September. of *SPE Middle East Intelligent Oil and Gas Symposium*.

Camilleri, L., El Gindy, M. and Rusakov, A. (eds.). 2016a. ESP real-time data enables well testing with high frequency, high resolution, and high repeatability in an unconventional well, volume *All Days*. of *SPE/AAPG/SEG Unconventional Resources Technology Conference*.

Camilleri, L., El-Gindy, M. Rusakov, A., Bosia, F., Salvatore, P. and Rizza, G. (eds.). 2016b. i Testing the Untestable Delivering Flowrate Measurements with High Accuracy on a Remote ESP Well, volume Day 2 Tue., 08 November. *Abu Dhabi International Petroleum Exhibition and Conference, 2016*.

Camilleri, L., El Gindy, M., Rusakov, A., Ginawi, I., Abdelmotaal, H., Sayed, E., Edris, T. and Karam, M. (eds.). 2017. Increasing production with high frequency and high-resolution flow rate measurements from ESPs, volume Day 2 Tue., 25 April *SPE Gulf Coast Section Electric Submersible Pumps Symposium, 2017*.

Carpenter, C. 2019. Analytics solution helps identify rod-pump failure at the wellhead. *Journal of Petroleum Technology* 71(05): 63–64.

Cramer, R. and Goh, K-C. (eds.). 2009. Data driven surveillance and optimisation for gas, subsea and multizone wells, volume *All Days*. *SPE Digital Energy Conference and Exhibition, 2009*.

Cramer, R., Griffiths, W.N., Kinghorn, P., Schotanus, D., Brutz, J.M. and Mueller, K. (eds.). 2011. Virtual measurement value during start up of major offshore projects, volume *All Days*. *IPTC International Petroleum Technology Conference, 2011*.

Denney, T., Wolfe, B. and Zhu, D. 2013. Benefit evaluation of keeping an integrated model during real-time ESP operations. *In*: *All Days, SPE Digital Conference, SPE*, pp. SPE–163704–MS.

Garcia, A., Almeida, I., Singh, G., Purwar, S., Monteiro, M., Carbone, L. and Herdeiro, M. 2010. An Implementation of On-Line Well Virtual Metering of Oil Production.

Goh, K-C., Dale-Pine, B., Yong, I.H.W., Peter, V. and Lauwerys, C. (eds.). 2008. Production surveillance and optimisation for multizone smart wells with data driven models, volume *All Days. SPE Intelligent Energy International Conference and Exhibition, 2008.*

Grimstad, B., Gunnerud, V., Sandnes, A., Shamlou, S., Skrondal, I.S., Uglane, V., Ursin-Holm, S. and Foss, B. (eds.). 2016. A simple data-driven approach to production estimation and optimization, volume *All Days. SPE Intelligent Energy International Conference and Exhibition, 2016.*

Haitao, Y. et al. 2014. Real time calculation of fluid level using dynamometer card of sucker rod pump well, volume *All Days. IPTC International Petroleum Technology Conference*, Dec. IPTC-17773-MS.

Haouche, M., Tessier, A., Deffous, Y. and Authier, J-F. (eds.). 2012. Virtual flow meter pilot: based on data validation and reconciliation approach, volume *All Days. SPE International Production and Operations Conference and Exhibition, 2012.*

Khan, M.K., Tariq, Z. and Abdulraheem, A. 2020. Application of artificial intelligence to estimate oil flow rate in gas-lift wells. *Natural Resources Research* 29(6): 4017–4029.

Krikunov, D., Kosyachenko, S., Lukovkin, D., Kunchinin, A., Tolmachev, R. and Chebotarev, R. (eds.). 2019. AI-based ESP optimal control solution to optimize oil flow across multiple wells, volume Day 2 Tue., 22 October. *SPE Gas, 2019.*

Law, H.Y., Phua, P.H., Briers, J. and Kong, J. (eds.). 2018. Extending virtual metering to provide real time exception based analytics for optimising well management and chemical injection, volume Day 2 Wed., 21 March. *Offshore Technology Conference Asia, 2018.*

Li, X., Gao, X., Cui, Y. and Li, K. 2013. Dynamic liquid level modelling of sucker-rod pumping systems based on Gaussian process regression. pp. 917–922. *In*: *2013 Ninth International Conference on Natural Computation (ICNC).*

Loh, K. and Omrani, P.S. 2018. *Deep Learning and Data Assimilation for Real-Time Production Prediction in Natural Gas Wells*. ArXiv, abs/1802.05141.

Manikonda, K., Hasan, A.R., Obi, C.E., Islam, R., Sleiti, A.K., Abdelrazeq, M.W. and Rahman, M.A. 2018. Application of machine learning classification algorithms for two-phase gas-liquid flow regime identification. *Paper presented at Abu Dhabi International Petroleum Exhibition and Conference. In*: Day 4 Thu., 18 November, SPE. D041S121R004.

Mask, G., Wu, X. and Ling, K. 2019. An improved model for gas liquid flow pattern prediction based on machine learning. *In*: Day 2 Wed., 27 March. *Paper presented at the International Petroleum Technology Conference, Beijing China,* IPTC. D021S026R005.

Olivares, G., Escalona, C. and Gimenez, E. 2012. Production monitoring using artificial intelligence, APLT asset. *Paper presented at SPE Intelligent Energy International Utrecht, The Netherlands*. March. *In*: *All Days*, SPE. pp. SPE–149594–MS.

Peng, Y., Xiong, C., Zhang, J., Zhang, Y., Gan, Q., Xu, G., Zhang, X., Zhao, R., Shi, J., Liu, M., Wang, C. and Chen, G. 2019a. Innovative deep autoencoder and machine learning algorithms applied in production metering for sucker-rod pumping wells, volume Day 1 Mon., 22 July *Unconventional Resources Technology Conference*. SPE/AAPG/SEG. D013S011R004.

Peng, Y., Zhao, R., Zhang, X., Shi, J., Chen, S., Gan, Q., Li, G., Zhen, X. and Han, T. (eds.). 2019b. IInnovative Convolutional Neural Networks Applied in Dynamometer Cards Generation. *Paper presented at the SPE Western Regional Meeting*, San Jose, California, USA, April 2019. doi: https://doi.org/10.2118/195264-MS.

Poulisse, H., Van Overschee, P., Briers, J., Moncur, C. and Goh, K-C. 2006. Continuous well production flow monitoring and surveillance. *Paper presented at SPE Intelligent Energy International Amsterdam, The Netherlands, 11–13 April.*

Sabaa, A., El Ela, M.A., El-Banbi, A.H. and Sayyouh, Mohamed H.M. 2022. Artificial Neural Network model to predict production rate of electrical submersible pump wells. *SPE Production & Operations*, pp. 1–10. Sept.

Takacs, G. 2018. Chapter 1 - Introduction. pp. 1–10. *In*: Gabor Takacs (ed.). *Electrical Submersible Pumps Manual* (Second Edition). Houston, Texas: Gulf Professional Publishing,

Vinogradov, D. and Vorobev, D. (eds.). 2020. Virtual flowmetering novyport field examples, volume Day 3, Wed., 28 October. *SPE Russian Petroleum Technology Conference.*

Zhang, S. and Tang, Y. 2008. Indirect measurement of dynamometer card of pumping unit. pp. 952–4955. *In*: *2008 7th World Congress on Intelligent Control and Automation.*

Zhu, D., Alyamkin, S., Sesack, L., Bridges, J. and Letzig, J. (eds.). 2016. Electrical submersible pump operation optimization with time series production data analysis, volume *All Days*. *SPE Intelligent Energy International Conference and Exhibition.*

Zhu, D., Luo, X., Zhang, Z., Li, X., Peng, G., Zhu, L. and Jin, X. 2021. Full reproduction of surface dynamometer card based on periodic electric current data. *SPE Production & Operations* 36(03): 594–603.

Zhu, K., Wang, L., Du, Y., Jiang, C. and Sun, Z. 2020. Deeplog: Identify tight gas reservoir using multi-log signals by a fully convolutional network. *IEEE Geoscience and Remote Sensing Letters, title=Prediction of Subsurface NMR T2 Distributions in a Shale Petroleum System Using Variational Autoencoder-Based Neural Networks* 17(4): 568–571.

Ziegel, P., Shirzadi, S., Wang, S., Bailey, R., Griffiths, P., Ghuwalela, K., Ogedengbe, A. and Johnson, D. 2014. A data-driven approach to modelling and optimisation for a North sea asset using real-time data. *Paper presented at the SPE Intelligent Energy Conference & Exhibition*, Utrecht, The Netherlands, April 2014. doi: https://doi.org/10.2118/167850-MS.

CHAPTER 6

Data-driven and Machine Learning Approach in Estimating Multi-zonal ICV Water Injection Rates in a Smart Well Completion

Daniel Asante Otchere[1,2,*] *and*
Mohammed Ayoub Abdalla Mohammed[3]

1. Introduction

The 'smart' or 'intelligent' well is considered one of the highly developed types of nonconventional wells. This statement refers to the cutting-edge versions of wells that have been introduced in recent years. The progressive nature of these wells is evident in their ability to gather and analyse data, monitor and control the production process, and adjust production in response to changing reservoir conditions. The term 'smart' or 'intelligent' has been coined to highlight the high level of automation and the use of advanced technology in these wells. A typical smart well features a customised completion with packers or sealing components that partition the wellbore with downhole sensors and pressure control valves fitted on the production tubing across separate reservoir intervals in a heterogeneous formation. Downhole sensors allow for continuous monitoring of temperature and pressure across the reservoir and

---

[1] Centre of Research for Subsurface Seismic Imaging, Universiti Teknologi PETRONAS, 32610, Seri Iskandar, Perak Daril Ridzuan, Malaysia.
[2] Institute for Computational and Data Sciences, Pennsylvania State University, University Park, PA, USA.
[3] Chemical and Petroleum Engineering, UAE University, Sheik Khalifa Street at Tawam R/A, Maqam District, Al Ain, United Arab Emirates.
* Corresponding author: ascotjnr@yahoo.com

control valves, which can be used to calculate approximately zonal flow rates. In contrast, downhole valves like Inflow Control Valves (ICVs) are flexible in controlling zonal flow rates and are used as a control variable in optimising wells to enhance recovery. A smart well can be multilateral with an ICV controlling each lateral or a single bore well with an ICV controlling each zone. Most of the new oil and gas field developments include smart wells. They achieve the desired output while lowering capital and operating costs.

Smart well completion is a technology used to optimise oil and gas reservoir production. It involves collecting, transmitting, and analysing completion, production, and reservoir data, allowing for remote selective zonal control. The utilisation of this technology contributes to the improvement of production and ultimate recovery, as well as the decrease in capital and operating expenditures. Smart well completion systems are meticulously designed to cater to the global demand for intelligent completions, even in the most difficult conditions. These systems are comprised of a variety of components, such as permanent monitoring and downhole control systems, zonal isolation and interval control devices, distributed temperature sensing systems, surface control and monitoring systems, data acquisition and management software, as well as system accessories. Multi-zonal reservoirs present a unique set of challenges for oil and gas production, and smart well completion technology is well suited to meet these challenges. By dividing the well completion into multiple production zones that can be controlled independently, intelligent well completion allows for selective control of production from different zones in the reservoir. This enables effective management of water injection, gas and water breakthrough, and individual zone productivity, which can help to increase ultimate recovery and reduce capital and operating expenditures.

Intelligent well completion technology can also improve the management of water injection and gas breakthroughs, which can reduce water production and increase ultimate recovery. By remotely controlling the ICVs, it is possible to manage the flow of fluids from different zones in the reservoir, leading to improved production efficiency and reduced costs. In addition to the benefits mentioned, intelligent well completion technology also allows for the reduction of the number of wells required for field development, which can lead to a decrease in drilling and completion costs. Furthermore, it allows for water management through remote zonal control, reducing surface handling facilities' size and complexity.

Operators of a multi-zone intelligent injection well face a problem in estimating and managing fluid distribution through zonal ICVs in each reservoir zone. Based on available wellbore, fluid, and well-string data, including the setting of the zonal ICVs, geometry size, production string, injection fluid properties, wellhead P/T data, and zonal reservoir pressure/injectivity data, a simple yet precise empirical correlation has been developed to estimate the volume of injection fluids to each reservoir zone. However, these equations can become out of sync and sometimes require recalibration. New downhole pressure regimes, new fluids (water/gas), and wellhead pressure are all situations that necessitate recalibration.

As a Production Technologist, one of the most important works is to estimate the oil production/water injection from/to multiple reservoir zones based on operational parameters such as bottomhole pressure, tubing differential pressure, and wellhead pressure. Most of these workflows or empirical correlations assume physics in some sort, which involves many assumptions. Hence, this research is being proposed to use a data-driven approach. This approach uses artificial intelligence (AI) techniques where the computer learns physics based on data patterns without assumptions and complex physics. This concept proposes advanced data analytics techniques, allowing new insights and understanding of the operational parameters (Otchere et al., 2022b). Advanced data analytics and machine learning will be used to build a model and predict water injection volumes into multiple reservoirs using daily production or injection data from an actual field. The main objectives of this work are to;

1. Study and understand the key operational parameters that are necessary to estimate zonal rates through the use of advanced data analytics techniques.
2. Create a reliable method using machine learning models to calculate production rates in multi-zonal reservoirs.

Given the success of AI in the oil and gas business, this initiative has tremendous research possibilities. Current research applies machine learning techniques to optimise ICV settings to increase production. In this project, eight machine learning models will be employed to estimate the injected water volumes into each reservoir when ICV settings are out of sync, giving wrong zonal estimates. This approach will replace recalibrating the empirical correlation to estimate injected water into multiple zones, resulting in interrupted field production. This approach will be reapplied in oil production wells with various input parameters upon acceptable results. This research could provide a simple and accurate data-driven approach to estimating each reservoir unit's volume of production fluids based on operational parameters. The method can also provide real-time estimates of produced volumes, aiding reservoir management plans. The current results submitted along this proposal could be extended to an advanced formulation of the reservoir management plan, daily production operational changes to meet critical targets and the operator's policy of developing domestic oil resources responsibly.

2. Brief Overview of Intelligent Well Completion

A smart well completion system improves production by gathering, transmitting, and analysing data related to completions, production, and reservoirs. This allows for remote control of specific zones, ultimately leading to increased efficiency in the reservoir. The system helps to boost production through the blending of production from different zones, enables better management of water injection and breakthroughs, and reduces both capital and operating expenses by minimising the number of wells needed and optimising production without costly interventions. The smart well system is designed to operate in various environments and can include a combination of devices for zonal isolation, interval control, downhole control, permanent monitoring, and temperature sensing, as well as software for data acquisition and management.

2.1 ICV Setting and Determination

Multiple production simulations should be conducted to optimise the design and sizing of ICVs. The ICV size has a significant impact on the production rate, which can vary widely across different fields. To achieve specific goals, such as maximising oil recovery or minimising water production, it is essential to determine the optimal configuration for the ICVs. This can be accomplished by designing the ICVs in a way that ensures each setting yields a unique production rate. This approach greatly influences the optimisation process, as the resulting data provides valuable insights into the optimal ICV configuration for achieving the desired outcome.

In the domain of ICV design, the process for determining the optimal design is usually contingent on the production capability of each lateral. However, in practice, several oil companies have streamlined the process by employing the average field production rate as a basis for design, rather than individual lateral rates. This approach is often deemed more expedient, particularly for logistical purposes, including maintenance and surveillance. For this research, ICV size was selected based on individual zonal flow at a particular setting, commingled flow from two zones iteratively and commingled flow from all three zones whiles monitoring downhole pressures. The ICV settings were discretised so that each setting resulted in production rates corresponding to zonal pressures. Several factors taken into account in the design of the ICV include:

1. Simulations were run to establish the minimum and maximum production rates for each zone where all zones were commingled and when a single zone was produced sequentially.

2. The discretisation of the ICV region was accomplished by establishing predefined intermediate settings (ranging from 0 to 10) that correspond

to the minimum and maximum production rates. The ICV region was segmented into these settings to provide a quantitative representation of the valve's opening, where 0 denotes complete closure and 10 denotes complete openness.

3. In accordance with downhole pressure monitoring, the maximum production rate can be achieved by adjusting the ICV (intelligent completion valve) to its highest setting range (7–10), whereas the minimum ICV setting range (0–3) results in a production rate of zero.
4. If production rates significantly differed between zones, different sized ICVs were used appropriately, although this may not be a viable technique.
5. The new ICV settings were tested on different base cases of all zones producing and one zone producing individually. If more than half of the ICV settings in a zone gave the same production rate, the maximum production rate was reduced. Also, the cases simulated were to capture instances where a particular zone experienced production issues (gas/water breakthroughs) and had to be shut in. Although the data was applied to an injection, this approach was simulated considering a production well to capture various scenarios and production issues.

2.2 Literature Review of ICV Innovations and Machine Learning Applications

In literature, ICVs have been widely studied as a means of optimising production in oil and gas reservoirs. These valves are used to manage the inflow of fluid into a well, allowing for selective control of production from different zones in the reservoir. This selective control enables effective management of water injection, gas and water breakthrough, and individual zone productivity, which can help increase ultimate recovery and reduce capital and operating expenditures.

AI has been successfully implemented in the petroleum industry, including reservoir characterisation (Otchere et al., 2021), seismic interpretation (Otchere et al., 2022c), reservoir engineering (Otchere et al., 2022a) and many other sectors. Machine learning techniques have also been applied to the optimisation of ICV valves. Researchers have used machine learning algorithms to model the relationship between ICV settings and production rates and to predict the optimal ICV settings for a given set of conditions. These models can take into account various factors such as wellbore geometry, fluid properties, and reservoir characteristics.

Various techniques have been suggested to enhance the productivity of smart wells. For example, Mubarak et al. (2008) employed an intensive production test to reduce water production from a trilateral smart well, where the laterals were found to be sensitive to ICV settings. Different settings were observed to be practical for different sources of water production. Similarly, Jalali et al. (1998) improved the delivery rate of a smart gas well by unchoking the bottom layer gradually after producing the top layer without any restriction. In addition, Yeten et al. (2002) employed a gradient-based approach to optimise the cumulative oil recovery from smart wells. Naus et al. (2006) put forward a methodology to determine the changes in flow rate due to ICV settings and tested it in two reservoirs with the objective of achieving maximum ultimate recovery. Furthermore, Brouwer and Jansen (2002) evaluated the impacts of smart completions with several well targets and limitations, while Alhuthali et al. (2010) used smart wells and optimal rate control for waterflood optimisation. Yeten (2003) proposed a general methodology that uses genetic algorithms, hill climbing, and artificial neural networks (ANNs) to optimise well type, trajectory, location, and ICV settings. Another study, conducted by Behrouz et al. (2016), is the use of an intelligent well technology to develop a new method for selecting and ranking candidate wells and fields, determining interval control valve (ICV) size, and optimising ICV settings. An efficient ICV setting optimisation in an intelligent well was proposed, which maximised cumulative oil, minimised water production or conducted both. Real case studies were considered to demonstrate the effectiveness and robustness of the proposed methodology, resulting in a considerable improvement in the objective function using the developed methodology.

Huang et al. (2011) aimed to develop an intelligent well completion system in China by optimising downhole monitoring, data transmission, intelligent optimisation, and production control systems. The authors developed an ICV system with proprietary internet protocol (IP) and created the first set of intelligent completion system (ICS) in China for measurable and optimisable oil well exploitation. Results showed that the system is stable and reliable, resulting in a 10.5% increase in oil production. Malakooti et al. (2020) proposed an integrated control and monitoring (ICM) algorithm that can enhance production from multi-zone I-wells. The algorithm employs a two-level optimisation process to maximise either projected zonal or oil production reliability. The authors validated the algorithm by subjecting it to a commercial transient wellbore simulator. The simulation results demonstrate that the ICM algorithm can optimise oil production while reducing the number of flow tests required. Furthermore, the study highlights that the exact estimation of zonal properties is not essential to achieve maximum oil production in multi-zone I-wells. The algorithm's efficacy suggests that it may be applicable to optimising multi-well flow rate allocation and the start-up process of multi-zone I-wells.

In literature, researchers have also used other machine learning techniques such as reinforcement learning, decision tree, and fuzzy logic to optimise the

ICV valves. These techniques have shown promising results in finding the optimal ICV settings and improving the performance of oil and gas reservoirs. Overall, the literature review suggests that combining ICV data and machine learning techniques can be an effective approach for optimising production in oil and gas reservoirs. Machine learning algorithms can be used to model the relationship between ICV settings and production rates and to predict the optimal ICV settings for a given set of conditions. This can help to increase ultimate recovery, reduce capital and operating expenditures, and improve the performance of oil and gas reservoirs.

3. Methodology

In the context of ICV design, the process usually relies on the production capacity of individual laterals. Notwithstanding, some oil companies opt for a simplified approach by utilising the mean production rate of the field rather than individual lateral rates.

1. *Data acquisition*: For this research, data is to be acquired from the production technology team in the subsurface department. The data required from multiple wells are SPFM WI Rate, Annulus Pressure, Tubing Pressure, ICV dP, ICV position, Annulus Temperature for the separate reservoir zones (lower, middle and upper), MPFM Rate, THT (tubing head temperature), Annulus Pressure, THP (tubing head pressure), Tubing dP (differential pressure). Having some ICV mappings in the wells of interest is also essential. The data from the mapping helps gain insight and train the machine learning model on accurate data. The data were grouped per hour and within a two-year range.

2. *Data preprocessing and preparation*: Based on the collected data, preprocessing techniques are used to clean the data. Missing values and data variables were removed to prepare the data for further analysis. The total number of data used, based on the descriptive statistics after the removal of missing data, is 3,767 data points.

3. *Exploratory data analysis*: The most critical and time-consuming part of any machine learning project is the exploratory data analysis. It is the first direct encounter with the data towards understanding it. Initial data investigations were carried out to discover patterns in data, spot anomalies, and check assumptions with the help of summary statistics, heatmaps, and pair plot visualisation.

4. *Explainable AI and feature selection*: In this step, the various input features are inspected and checked for their importance. Based on the results from the data analysis, feature selection is to be performed using explainable AI techniques (model agnostic metrics). The approach is

different from the standard methods that operate on correlation. Since correlation does not mean causation, the model agnostic metric is used to indicate the variables that have a causal effect on estimating zonal rates.

5. *Machine Learning*: Several advanced machine learning models were used to estimate the zonal rates for each reservoir from the data provided. A holdout cross-validation of 10% was applied to segregate the data into training (3,390), and test (377) sets. The performance of these models will be compared using several error metrics and visualisation techniques. The model with the least errors, high accuracy, and high probability of fit will be selected and deployed.

The aim is to assess when the empirical correlation goes out of sync and if the model performance deteriorates over time, assuming all conditions remain the same. A suitable machine learning model that can estimate zonal rates can be used to estimate the required injection volumes needed for specific zones. This model can help production technologists plan the optimum ICV positions and tubing head injected volumes required to sustain pressures in the relevant zones. This approach will also limit the need for strategic ICV mapping during pressure build up (PBU) and delays in bringing production up to the required optimal level before shut-in. A schematic of the workflow is illustrated in Fig. 6.1.

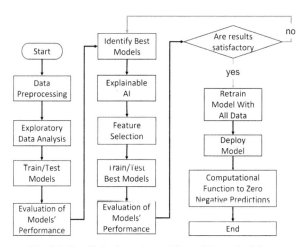

Fig. 6.1. Detailed schematic workflow of the methodology.

3.1 A Brief Overview of Models Used in This Study

The summary analysis of the reviewed models used in estimating zonal rates in this study is summarised in Table 6.1.

Table 6.1. Summary of models used in this study.

Base learner	Description	Developed or Implemented by Authors
Bootstrap aggregation (Bagging)	Bootstrap aggregation of trees is used to reduce the variance for algorithms that have high variance.	Breiman (1996)
Ridge Regression	Ridge regression is a robust technique against multicollinearity. The algorithm minimises standard errors by applying some bias to the model estimates, resulting in a more reliable prediction. The algorithm calculates the difference between the means of the standardised dependent and independent variables and dividing by their standard deviations.	Kuhn and Johnson (2013)
Least absolute shrinkage and selection operator (Lasso) Regression	Adds a penalty term to the regression equation that forces the sum of the absolute values of the regression coefficients to be less than a specific value. This has the effect of shrinking the coefficients of less important variables to zero, effectively removing them from the model.	Tibshirani (1996)
KNeighbors Regressor	It is based on the idea that data points with similar features tend to have similar output values. The KNN regressor predicts the output value for a new data point by looking at the K-nearest training data points and averaging their output values. The value of K determines the number of neighbours considered in the prediction, and the distance metric used to measure the similarity between data points can be chosen based on the problem at hand.	Silverman and Jones (1989)
Decision Tree	It uses induction and pruning techniques to build hierarchical decision boundaries and remove unnecessary structures from the decision tree to battle overfitting.	Gordon et al. (1984)
Extreme Gradient Boosting (XGBoost)	An implementation of gradient boosted decision trees of the first and second-order to maximise the loss function adding an extra regularisation term to adjust the final weights acquired to avoid overfitting.	Chen and Guestrin (2016)
Extremely Randomised Trees (Extra tree)	Potentially performs better than the random forest, although it uses a simpler algorithm to construct the decision trees used as ensemble members.	Geurts et al. (2006)
Random Forest	An advanced decision tree that is robust against overfitting and offers easy interpretability.	Breiman (2001)

3.2 Criteria for Model Evaluation

Successfully training a model and achieving a good convergence does not signify the end of the model process. In model regression analysis, the model's error is the difference between the actual data points and the best fit line produced by the algorithm. There is not a single statistic that can analyse all forms of data. The metrics are impacted by a number of components, including but not limited to the presence of outliers, the machine learning algorithm used, the ease with which derivatives can be discovered, and the prediction confidence. Therefore, the model errors were evaluated using the following standards to determine whether the models utilised in this research were appropriate;

1. *Mean Absolute Error (MAE)*: The MAE is a well-established evaluation metric used to quantify the accuracy of predictive models. It is defined as the L1 loss, representing the sum of absolute differences between the predicted output and the target variable. By measuring the average magnitude of errors without taking into account the direction of errors, the MAE provides a reliable measure of model performance. Additionally, this metric is particularly sensitive to relative errors, making it a useful tool for assessing performance in applications where errors of all sizes are equally important. The MAE is also robust against global scaling of the predicted output and the presence of outliers, making it a versatile metric for evaluating model performance in various settings. Mathematically, the MAE is written as;

$$MAE = \frac{\sum_{i=1}^{n}|y_i - \hat{y}_i|}{n} \qquad (6.1)$$

2. *Root Mean Squared Error (RMSE)*: This performance evaluation metric is popular because it is interpretable as the standard deviation of the model's prediction errors and specifies the closeness of predicted data to actual data. This function is written as follows;

$$RMSE = \sqrt{\frac{\sum_{i=1}^{n}(y_i - \hat{y}_i)^2}{n}} \qquad (6.2)$$

3. *Akaike Information Criterion (AIC)*: This evaluation standard was developed using a frequentist probability framework that assigns a model a score based on its maximum likelihood estimation. This technique assesses models' quality and accuracy, resulting in a better model fit of the data. This criterion is expressed as;

$$AIC = 2K - 2(log - likelihood) \qquad (6.3)$$

In order to assess the accuracy of prediction models based on the number of input features, various evaluation metrics are employed. However, except for R^2, a lower value for the performance evaluation criteria indicates better model performance. If a conclusion cannot be drawn based on these parameters, a ranking rule is employed to determine precedence. Specifically, the ranking order is established as follows: MAE, RMSE, AIC, and R^2.

4. Results and Discussion

4.1 Explainable AI

All 21 input features were used to train eight models, and their results are illustrated in Fig. 6.2. From the results, the worst performing models are the Lasso and Ridge algorithms. This result indicates that linear models are inappropriate for this study due to the high dimensional input data.

The RF model was used for the multi-output prediction of the smart well injection into the upper, middle, and lower reservoir under study. The same training and testing datasets used in training all the models were used in this approach. The Permutation Feature Importance (PFI) analysis was used to identify the relevant features that can help predict the volume of water being injected in multiple zones. Model agnostic metrics are useful since machine learning models are designed to be interpretable to help understand how a model generated a given prediction.

Figure 6.3 visualises the PFI values as absolute values for both train and test data. Many input variables have a significantly low importance score from the results. This result implies that a small number of variables can capture the predictive value of these input variables. The PFI plot demonstrates that

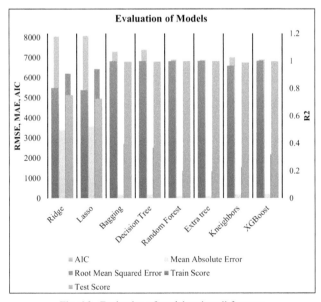

Fig. 6.2. Evaluation of models using all features.

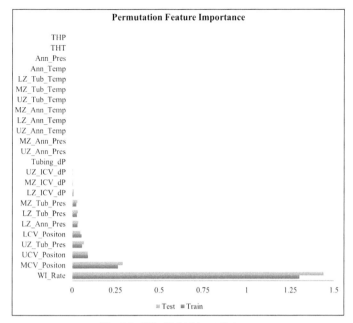

Fig. 6.3. PFI of initial input features.

permuting a variable reduces the multi-zonal prediction accuracy. According to this analysis, some of the characteristics are more significant than others. It is essential to state that this analysis was done for both the training and testing dataset to help account for features that may help with the generalisation power of the model. It was noticed that the top essential features are the same for both training and testing datasets. Features such as WI_Rate, the zonal ICV position, zonal annulus temperature, the tubing, and ICV differential pressures were considered essential. This result suggests that these features will contribute to the generalisation power of the model. If a feature is deemed necessary for the train set but not for the testing, this feature will probably cause the model to overfit. An example of this feature is the UZ_ICV_dP. Again, when all logs, including the new features, were inputted, the mean test accuracy was 0.998.

As such, sensitivity analysis were performed to identify the number of input features relevant to this study. Using the six high-performing models above, the top 10 and seven features were analysed against all features using accuracy. From the results in Fig. 6.4, the models scored high accuracy when seven input features were used. The features exhibiting causal relations to the three targets are WI_Rate, UZ (upper zone)_Tub_Pres, MZ (middle zone) _Tub_Pres, LZ (lower zone)_Tub_Pres, UCV_(upper control valve) Position, MCV (middle control valve)_Position, LCV (lower control valve)_Position.

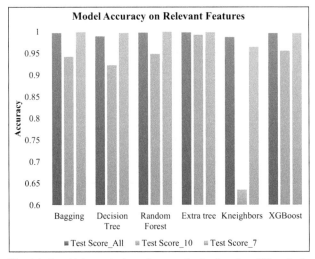

Fig. 6.4. Sensitivity analysis on feature selection based on PFI analysis.

Per domain knowledge, these features relate to the total injected volume, the zonal ICV position, which indicates that a particular zone is closed or open, and the tubing pressures of each zone.

4.2 Model Evaluation

Further evaluation was conducted on the seven features selected as relevant to identify the top-performing models for further optimisation. Significantly, the ET and RF models exhibited superior performance compared to the other models, even though none of them demonstrated overfitting when evaluated on the holdout data. As a result, we conducted additional analysis to assess the prediction errors of the models. Figure 6.5 illustrates the accuracy and

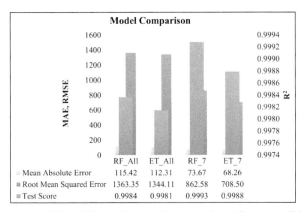

Fig. 6.5. Comparison of RF and ET models using the seven relevant features to all input features.

consistency of all the recommended models, based on the error measurements. MAE, RMSE, and R^2 results indicate that the seven features resulted in massive improvements in model performance. Although the RF model achieved a higher test accuracy, the ET model performed better comparatively in terms of MAE and RMSE. In a broader perspective, the seven features resulted in a 40% and 47% reduction in the MAE and RMSE, respectively.

4.3 Sensitivity Analysis

Based on the test data, Figs. 6.6 to 6.8 show the kernel density estimation (KDE) of the expected and actual zonal rates. The actual values are depicted in red, while the expected values are indicated in blue. The zonal injection rates predicted by the ET model are much closer to the observed test data. This suggests that the ET model is capable of accurately representing the diverse range of the data, which makes it appropriate for use in predicting zonal injection rates for other wells in the same field. In addition, the assessment metrics discussed earlier are confirmed by the sensitivity analysis.

Figures 6.9 to 6.11 show the joint plot and regression combination to visualise the predicted zonal and actual test data zonal rates and error margins. The joint plot function is a distribution plot that shows univariate histogram KDE curves and bivariate graphs. Analysis of the regression plot and 95% confidence interval illustrated in these figures indicate that the ET model using the seven PFI selected features performed accurately. The ET model showed a higher confidence interval and matched the range of injected volumes based on the histogram and KDE plot. However, the confidence

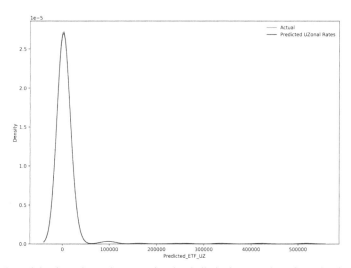

Fig. 6.6. Kernel density estimate demonstrating the similarity between the estimated and actual test upper zonal rates.

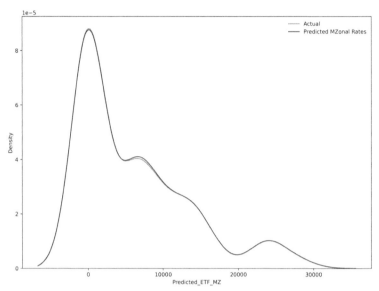

Fig. 6.7. Kernel density estimate demonstrating the similarity between the estimated and actual test middle zonal rate.

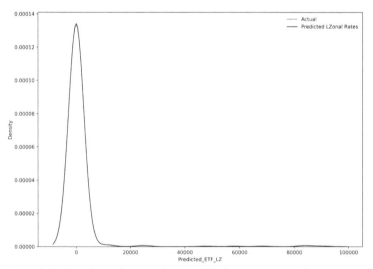

Fig. 6.8. Kernel density estimate demonstrating the similarity between the estimated and actual test lower zonal rate.

level tends to deteriorate as high volumes of water were predicted for the upper zone, which is analysed further. For the ET model, most data after 200,000 bbls (barrels) were widely scattered due to insufficient high injected volume data into that zone. The ET model, however, successfully captured the

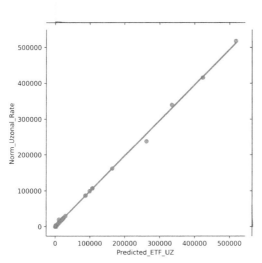

Fig. 6.9. Cross-plot of ET predicted vs actual upper zonal rates.

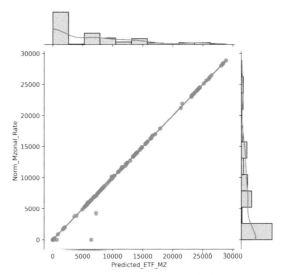

Fig. 6.10. Cross-plot of ET predicted vs actual middle zonal rates.

trends despite insufficient data from this analysis. Although the lower zone experienced similar data inadequacy in high injected volumes, the model did not experience low confidence in its prediction. This observation can be attributed to the variation in the training data and the wide spread of test data for the lower zone compared to the upper zone with a high concentration of low injection volumes.

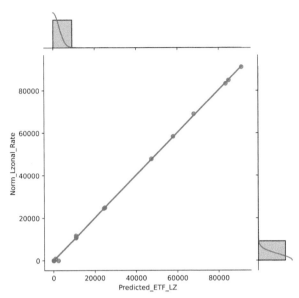

Fig. 6.11. Cross-plot of ET predicted vs actual lower zonal rates.

4.4 Model Deployment

The excellent results achieved for the multi-zonal water injection rates by the ET model using the model agnostic selected features gave confidence in its deployment. Data from another well with calculated ICV zonal rates when the ICV curves used to determine the rates were in sync was used. The data had 3,767 data points representing about five months of hourly data. Figure 6.12 shows the results of the predicted rates for the injected well the model was deployed on. The accuracy of the combined zonal rate prediction to the actual water injection is apparent and confirms the accuracy of the model. The zonal rates, illustrated in Figs. 6.13 to 6.15, will show the predicted rates and how

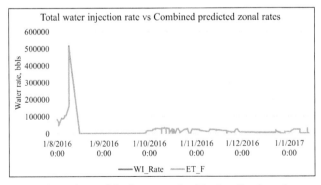

Fig. 6.12. Total water injection rate vs Combined predicted zonal rates.

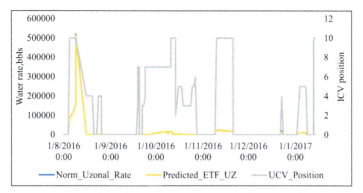

Fig. 6.13. Comparison of actual and predicted upper zonal ICV rates against ICV position.

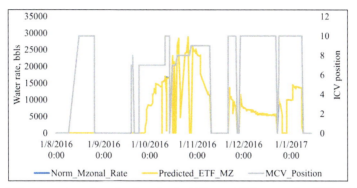

Fig. 6.14. Comparison of actual and predicted middle zonal ICV rates against ICV position.

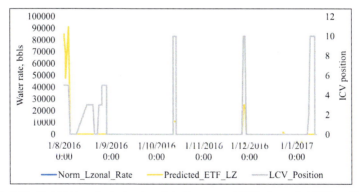

Fig. 6.15. Comparison of actual and predicted lower zonal ICV rates against ICV position.

they make predictions based on ICV positions. This is to confirm that all the predicted zonal rates went to the actual zone where the ICV was opened.

5. Conclusions

In summary, eight supervised machine learning models were employed to predict multi-zonal rates for a smart water injection well in an oil and gas production setting. The machine learning models were trained and evaluated using PFI, a model-agnostic metric, to identify relevant features. Seven of the features were relevant and used as input to further train the top six models from the previous evaluation where all features were used. The Extra Trees model achieved the highest precision and consistency of 708 bbls RMSE and 68 bbls MAE compared to the test data. The Extra Trees predicted zonal rates were also visualised using KDE and joint plots to confirm their accuracy. Upon satisfactory results, the Extra Trees model was deployed on a new well with five months of hourly data, and the combined predicted zonal rates matched the total injected rate. Additionally, the model could predict zonal rates in instances where the ICVs were in the closed and fully opened positions. Overall, the results of this research demonstrate the potential of machine learning in predicting multi-zonal rates in oil and gas production and highlight the use of the Extra Trees model as a robust and effective tool for this task.

This research provides preliminary results for the approach employed in water injection wells. When used in producer wells, this method could provide a simple and accurate data-driven approach to estimating each reservoir unit's volume of produced fluids based on downhole parameters. The method can also provide real-time estimates of produced volumes, aiding advanced formulation of a reservoir management plan, daily production operational changes to meet critical targets, and the operator's policy of developing domestic oil resources responsibly. This project's success will not end ICV mapping but will drastically reduce the number of mappings performed in the life of the field. One limitation of this study is that the model is as good as the data used to train it. When new data is introduced that falls out of the range of the model, ICV mapping will be required, and the model will be retrained with the new data to fine-tune its prediction. This limitation, however, does not downplay the immense advantages this study provides. This approach is recommended to save time and money during production using a smart well completion.

Code Availability

The Jupyter Notebook used in this study is hosted at https://github.com/ascotjnr/Smart-Well-Completion.

Acknowledgement

The authors wish to thank the University Teknologi Petronas and the Centre for Subsurface Seismic Imaging and Hydrocarbon Prediction for supporting this work.

References

Alhuthali, A.H., Datta-Gupta, A., Yuen, B. and Fontanilla, J.P. 2010. Field applications of waterflood optimisation via optimal rate control with smart wells. *SPE Reservoir Evaluation & Engineering* 13: 406–422. https://doi.org/10.2118/118948-PA.

Behrouz, T., Rasaei, M.R. and Masoudi, R. 2016. A novel integrated approach to oil production optimisation and limiting the water cut using intelligent well concept: using case studies. *Iranian Journal of Oil and Gas Science and Technology* 5: 27–41. https://doi.org/10.22050/IJOGST.2016.13827.

Breiman, L. 1996. Bagging predictors. *Mach Learn.* 24: 123–140. https://doi.org/10.1007/bf00058655.

Breiman, L. 2001. Random forests. *Mach. Learn.* 45: 5–32. https://doi.org/10.1023/A:1010933404324.

Brouwer, D.R. and Jansen, J.D. 2002. Dynamic optimisation of water flooding with smart wells using optimal control theory. *In*: European Petroleum Conference. SPE, Aberdeen. https://doi.org/10.2118/78278-MS.

Chen, T. and Guestrin, C. 2016. XGBoost: A scalable tree boosting system. pp. 785–794. *In*: Proceedings of the ACM SIGKDD International Conference on Knowledge Discovery and Data Mining. ACM, New York, NY, USA. https://doi.org/10.1145/2939672.2939785.

Geurts, P., Ernst, D. and Wehenkel, L. 2006. Extremely randomised trees. *Mach. Learn.* 63: 3–42. https://doi.org/10.1007/s10994-006-6226-1.

Gordon, A.D., Breiman, L., Friedman, J.H. Olshen, R.A. and Stone, C.J. 1984. Classification and regression trees. *Biometrics* 40: 874. https://doi.org/10.2307/2530946.

Huang, Z., Li, Y., Peng, Y., Shen, Z., Zhang, W. and Wang, M. 2011. Study of the intelligent completion system for liaohe oil field. *Procedia Eng.* 15: 739–746. https://doi.org/10.1016/j.proeng.2011.08.138.

Jalali, Y., Bussear, T. and Sharma, S. 1998. Intelligent completion systems: the reservoir rationale. *In*: All Days. SPE. https://doi.org/10.2118/50587-MS.

Kuhn, M. and Johnson, K. 2013. *Applied Predictive Modeling*. New York, NY: Springer New York, https://doi.org/10.1007/978-1-4614-6849-3.

Malakooti, R., Ayop, A.Z., Maulianda, B., Muradov, K. and Davies, D. 2020. Integrated production optimisation and monitoring of multi-zone intelligent wells. *J. Pet. Explor. Prod. Technol.* 10: 159–170. https://doi.org/10.1007/s13202-019-0719-5.

Mubarak, S.M., Pham, T.R., Shammani, S.S. and Shariq, M. 2008. Case study: the use of downhole control valves to sustain oil production from the first maximum reservoir contact, multilateral, and smart completion well in ghawar field. *SPE Production & Operations* 23: 427–430. https://doi.org/10.2118/120744-PA.

Naus, N.M.J.J., Dolle, N. and Jansen, J.-D. 2006. Optimisation of commingled production using infinitely variable inflow control valves. *SPE Production & Operations* 21: 293–301. https://doi.org/10.2118/90959-PA.

Otchere, D.A., Arbi Ganat, T.O., Gholami, R. and Lawal, M. 2021. A novel custom ensemble learning model for an improved reservoir permeability and water saturation prediction. *J. Nat. Gas Sci. Eng.* 91: 103962. https://doi.org/10.1016/j.jngse.2021.103962.

Otchere, D.A., Abdalla Ayoub Mohammed, M., Ganat, T.O.A., Gholami, R. and Aljunid Merican, Z.M. 2022a. A novel empirical and deep ensemble super learning approach in predicting reservoir wettability via well logs. *Applied Sciences* 12: 2942. https://doi.org/10.3390/app12062942.

Otchere, D.A., Ganat, T.O.A., Nta, V., Brantson, E.T. and Sharma, T. 2022b. Data analytics and Bayesian Optimised Extreme Gradient Boosting approach to estimate cut-offs from wireline logs for net reservoir and pay classification. *Appl. Soft Comput.* 120: 108680. https://doi.org/10.1016/j.asoc.2022.108680.

Otchere, D.A., Tackie-Otoo, B.N., Mohammad, M.A.A., Ganat, T.O.A., Kuvakin, N., Miftakhov, R., Efremov, I. and Bazanov, A. 2022c. Improving seismic fault mapping through data conditioning using a pre-trained deep convolutional neural network: A case study on Groningen field. *J. Pet. Sci. Eng.* 213: 110411. https://doi.org/10.1016/J.PETROL.2022.110411.

Silverman, B.W. and Jones, M.C. 1989. E. Fix and J.L. Hodges (1951): An important contribution to nonparametric discriminant analysis and density estimation. *International Statistical Review* 57: 233–247.

Tibshirani, R. 1996. Regression shrinkage and selection via the lasso. *Journal of the Royal Statistical Society. Series B (Methodological)* 58: 267–288.

Yeten, B., Durlofsky, L.J. and Aziz, K. 2002. Optimisation of smart well control. *In*: SPE *International Thermal Operations and Heavy Oil Symposium and International Well Technology Conference*. Calgary.

Yeten, B. 2003. *Optimum Deployment of Nonconventional Wells* (PhD). Stanford University, California.

CHAPTER 7

Carbon Dioxide Low Salinity Water Alternating Gas (CO$_2$ LSWAG) Oil Recovery Factor Prediction in Carbonate Reservoir
Using Supervised Machine Learning Models

Eric Thompson Brantson,[1,*] *Zainab Ololade Iyiola,*[1]
Yao Yevenyo Ziggah,[2] *Alexander Ofori Mensah,*[1]
Daniel Asante Otchere,[3,4] *Efua Eduamba Abakah-Paintsil*[1]
and *Emmanuel Karikari Duodu*[1]

1. Introduction

With rising global energy demand and diminishing oil reserves, enhanced oil recovery (EOR) from existing brownfields is becoming increasingly important (Sheng, 2011). As production from the petroleum reservoir increases, the reservoir's inherent primary energy depletes, resulting in insufficient reservoir pressure to drive the oil to the surface. As a result, secondary recovery methods (water or gas injection) are needed to boost production output. The oil recovered through both primary and secondary processes ranges from about

---

[1] Department of Petroleum and Natural Gas Engineering, School of Petroleum Studies, University of Mines and Technology, Tarkwa, Ghana.
[2] Department of Geomatic Engineering, Faculty of Geosciences and Environmental Studies, University of Mines and Technology, Tarkwa, Ghana.
[3] Centre of Research for Subsurface Seismic Imaging, Universiti Teknologi PETRONAS, 32610, Seri Iskandar, Perak Daril Ridzuan, Malaysia.
[4] Institute for Computational and Data Sciences, Pennsylvania State University, University Park, PA, USA.
* Corresponding author: etbrantson@umat.edu.gh

20%–40% of the original oil in place (OOIP) (Stalkup, 1984). Following the application of primary and secondary oil recovery techniques, two-thirds of the OOIP remains in the reservoir (Gbadamosi et al., 2019).

According to Brantson et al. (2019), tertiary oil recovery methods are capable of retrieving a greater amount of oil than primary or secondary recovery methods. In this study, EOR synergy methods involving the injection of low salinity water (LSW) and carbon dioxide (CO_2) gas were employed to enhance oil recovery at the field scale. The use of LSW in field trials has demonstrated a significant improvement in oil recovery (Robertson, 2007; Webb et al., 2004; Vledder et al., 2010), making it a favourable option over conventional chemical EOR methods in terms of chemical costs, environmental impact, and field process implementation (Dang et al., 2014). On the other hand, CO_2-EOR has the potential to recover an additional 15%–20% of the remaining oil in place (Ahmed, 2018). The two main mechanisms behind CO_2 injection schemes are oil swelling and viscosity reduction (AlQuraishi et al., 2019). It is noted that all reservoir lithologies, including siliciclastic, carbonate, and others, are appropriate for CO_2-EOR provided they have an enough seal to contain hydrocarbons and interconnected pore space for fluid accumulation and flow (Verma, 2015). Therefore, estimating hydrocarbon's ultimate recovery factor provides more detailed insights into oilfield development strategies (Roustazadeh et al., 2022). The methods of material balance, decline curve analysis, and dynamic numerical simulation are used to estimate recovery factors, but they are time-consuming.

Over the years, there has been increasing awareness and concern over the continuous buildup of greenhouse gases in the atmosphere. This is due to human activities such as burning of fossil fuels, deforestation, and industrial processes that emit greenhouse gases such as carbon dioxide (CO_2), methane, and nitrous oxide. The accumulation of these gases has led to an increase in global temperatures, causing significant changes in the climate, which poses a threat to the entire world. One of the consequences of climate change is rising sea levels, which have been observed to be a direct result of the increasing levels of CO_2 and other greenhouse gases in the atmosphere. The rising sea levels have led to devastating consequences, such as flooding of coastal areas, displacement of communities, loss of habitats, and destruction of infrastructure (Santos et al., 2014). Therefore, it is crucial to find ways to mitigate the impact of greenhouse gas emissions on the environment and the world at large.

In recent years, there has been considerable attention given to the potential use of CO_2-EOR as a long-term anthropogenic CO_2 storage application (Mandadige et al., 2016). CO_2-EOR is a technique used in the oil and gas industry to extract more oil from wells by injecting CO_2 into the reservoir. This technique not only increases the amount of recoverable oil but also has the added benefit of storing the CO_2 underground. This can be an effective way to mitigate the impact of greenhouse gas emissions on the environment and provide a solution to the problem of global warming. Therefore, it is important for researchers, policymakers, and stakeholders to continue exploring the potential of CO_2-EOR as a long-term anthropogenic CO_2 storage application, while also working towards reducing greenhouse gas emissions and finding sustainable solutions to climate change.

The dominant cause of observed anthropogenic global warming has been unrestricted CO_2 emissions from fossil fuel combustion due to the rising usage of fossil fuels in producing electricity and other manufacturing processes (Kharecha et al., 2008). Over the last 35 years, the oil and gas industry has made numerous technological advancements and operational practices for injecting CO_2 for enhanced oil recovery (Yong et al., 2016). CO_2 flooding has emerged as one of the most promising EOR methods because it uses readily available, naturally occurring CO_2 from reservoirs (Sun et al., 2017). Typical incremental oil recovery by CO_2 flooding ranges between 5%–25% (Wu et al., 2021). The CO_2 injection method has traditionally been used in reservoirs with an oil gravity of less than 25 (Stosur, 2003).

In the quest to find the optimal use of CO_2 in enhanced oil recovery for secondary and tertiary modes, several techniques have been employed. These techniques include miscible and immiscible CO_2 flooding, CO_2 huff-n-puff, and CO_2-foam injection, among others (Christensen et al., 1998). Amongst these techniques, the combined forms of EOR methods have proven to optimise significantly oil recovery synergistically (Afzali et al., 2018; Teklu et al., 2016). A lot of CO_2-LSWAG experiments (AlQuraishi et al., 2011; Dang et al., 2014; Naderi and Simjoo, 2019; Pourafshary and Moradpour, 2019; Zolfaghari et al., 2013) in both carbonate and sandstone reservoirs have been performed which reported an incremental oil recovery from the initial oil in place but not much on reservoir scale.

Hybrid low-salinity gas flooding has garnered significant attention from researchers in recent years (Jiang et al., 2010; Kulkarni et al., 2004) due to its many advantages, such as low cost, low minimum miscibility pressure, and environmental friendliness. CO_2 is frequently used as the injection gas in this hybrid approach because of these benefits (Ma and James, 2022). The injection of LSW changes the gas solubility in water, affecting gas/oil interactions and ultimately enhancing oil recovery. Despite the potential benefits, there are varying observations regarding the effectiveness of this hybrid method, with some researchers reporting improved recovery. Others, however, have found no improvement in recovery over continuous gas injection (CGI). Furthermore, CO_2 partitioning between oil and water in the reservoir (Dang et al., 2014) can lead to different recovery rates depending on the reservoir conditions.

Low salinity CO_2-EOR was extensively described in the works of Zekri et al. (2015) for carbonate cores and Teklu et al. (2016) in carbonate cores and low permeability sandstone cores. Numerous studies on CO_2 LSWAG have been conducted in sandstone reservoirs with strongly water-wet sandstone cores and little clay content, which are unfavourable conditions for the LSW process (Dang et al., 2014) unlike carbonate reservoirs which are complex and have received relatively little attention. It was reported that by combining the low-salinity waterflooding method and the CO_2-EOR method, a new LSW-alternating-CO_2 (LS-WACO$_2$) EOR method emerges. It was concluded that the solubility of CO_2 in LSW was the main reason for the increased residual oil mobilisation more than in ordinary WAG. In the work of Dang et al. (2014), a hybrid LSW alternating miscible CO_2 flooding injection was modelled in a 1D core. Their research indicates that the hybrid approach is effective in resolving late production problems associated with water alternating gas WAG injection. Pourafshary and Moradpour (2019) conducted a comprehensive review of Dang et al.'s (2014) work and recommended further field-scale simulation studies based on experimental work to fully evaluate the useful benefits of the hybrid methods developed by Dang et al. (2014). Their recommendation was based on the promising results obtained by Dang et al. (2014), and highlights the need for more comprehensive studies to advance the understanding of this novel method. Furthermore, only very limited simulation studies have been done with regard to CO_2 LSWAG injection in carbonate reservoirs (Al-Shalabi et al., 2014; Hamoud and Pranoto, 2016).

In 2011, Aleidan and Mamora (2011) conducted a study that aimed to compare the effects of various water/CO_2 injection schemes on oil recovery from carbonate core samples. The study analysed the impact of SWAG and WAG injection schemes at different salinities. The researchers discovered that when they switched to LSW, there was an increase in oil recovery for both injection methods. This increase was as a result of the hybrid method and the synergy between gas and LSW injections. Gas dissolution in brine led to a reduction in fluid mobility, which accounted for the incremental oil recovery. Jiang et al. (2010) also conducted a study to investigate the impact of water salinity on the performance of WAG during miscible flooding in sandstones using highly viscous crude oil. The study aimed to examine the effect of low-salinity water on the effectiveness of WAG. The researchers discovered that in LSW, a higher gas solubility regulated the mobility ratio of water and viscous oil. This regulation occurred because the gas solubility in brine increases as salinity decreases, leading to a reduction in gas mobility and an increase in the viscosity of the brine. Consequently, the mobility of water and viscous oil became more balanced, which resulted in improved oil recovery rates. In addition to the findings mentioned previously, the research conducted by the team also showed that LSW is more effective than high-salinity water in improving oil recovery rates during high viscous crude oil sandstones miscible flooding. This is a critical finding that highlights the importance of considering the impact of water salinity on WAG performance during miscible flooding. The team's results demonstrate that the use of LSW can significantly increase oil recovery rates, providing further evidence of the potential benefits of LSW in enhanced oil recovery techniques. These findings have important implications for the petroleum industry, as they provide valuable insights into how water salinity can affect WAG performance, and how this knowledge can be used to optimise oil recovery rates in reservoirs with highly viscous crude oil.

In their experimental study, Al-Abri et al. (2019) investigated the effect of hybrid injection of immiscible CO_2 and smart water on sandstone core samples. The study found that the synergy between gas injection and the various ions present in the water samples led to a significant improvement in oil recovery. To conduct the study, the researchers utilized three synthetic brines, each containing 5000 ppm of $MgCl_2$, NaCl, and KCl, respectively. The research revealed that while $MgCl_2$-containing water had the highest solubility of CO_2 in brine, it resulted in the lowest oil recovery among the three tests. Furthermore, the multicomponent ion exchange of the smart water altered the rock's wettability, making it more water-wet, thus improving oil recovery without the need for gas injection.

The effectiveness of hybrid injection of LSW and miscible CO_2 for EOR in carbonates was investigated through a simulation study by Al-Shalabi et al. (2016) using the UTCOMP reservoir simulator. The study aimed to compare the performance of CGI with that of the hybrid method. The results showed that the CGI method achieved a high recovery of 98.9%, while the hybrid approach only increased it to 99.7% by controlling viscous fingering. Therefore, the study concluded that the hybrid method may not be suitable for conditions where gas miscibility is the primary mechanism for EOR. The study's findings could be useful in optimising the selection of EOR techniques for carbonate reservoirs to achieve maximum recovery with minimum effort and cost.

Consideration of the initial rock wettability is critical in hybrid LSW/gas methods. The alteration of wettability from oil-wet to water-wet is particularly significant in sandstone, as it positively affects the performance of LSW. Conversely, if the rock is initially water-wet, the LSW and hybrid methods will not be effective. Ramanathan et al. (2015) conducted an experimental study of seawater alternating gas (SeaWAG) and LSWAG injection for oil recovery from water-soaked sandstone. The study found that LSWAG had a lower recovery factor than SeaWAG due to the initial high water saturation of the rocks. In contrast, recovery by WAG increased from 76% to over 97% in an aged oil-wet core when low-salinity brine replaced seawater.

AlQuraishi et al. (2017) conducted a study that found that the low-salinity alternating miscible CO_2 method was not effective for clay-free sandstones, but did result in a recovery value of 35.1% of the OOIP when clays were present. Yang et al. (2005) conducted research and discovered that CO_2 has the ability to decrease the oil and brine interfacial tension (IFT) under constant temperature and pressure conditions. This decrease in IFT can contribute to additional oil recovery through the hybrid method. Further studies conducted by Teklu et al. (2016) and Ramanathan et al. (2016) showed that the decrease in IFT was less than 10 dynes/cm. However, a study by Kumar et al. (2016) reported comparatively high IFT in the presence of CO_2. Despite these mixed findings, it is important to note that the change in IFT is relatively small and is not considered to be the primary mechanism in the hybrid LSW/gas approach for increasing oil recovery. Other mechanisms, such as the reduction of residual oil saturation and improvement of displacement efficiency, are believed to play a more significant role in the hybrid method. Nonetheless, the discovery of the potential of CO_2 to reduce IFT remains an important aspect of research in the oil and gas industry.

The issue associated with pure gas flooding is unfavourable mobility resulting in viscous fingering and a reduction in volumetric sweep efficiency. With less gas needed for EOR projects, the WAG approach helps to solve this significant problem. Additionally, conventional CO_2 WAG methods typically cause a delay in oil production, which the current study can alleviate with CO_2 LSWAG. By overcoming the issue of late production that conventional WAG commonly faces, CO_2 LSWAG speeds up the synergy of these several process mechanisms. According to previous studies (Dang et al., 2014; Pourafshary and Moradpour, 2019; Sheng, 2014), the primary oil recovery mechanism in CO_2-LSWAG has been proposed to be wettability alteration to a more water-wet condition.

However, the modeling studies of CO_2-LSWAG in a 1D homogeneous model and at field scale in sandstone reservoirs (Dang et al., 2014) had been done and characterised by the expensive computational cost (Belazreg and Mahmood, 2020; Jaber et al., 2019). There is little or no application of fast and reliable machine learning models to forecast the performance of CO_2 LSWAG as shown in Table 7.1, for secondary and tertiary modes. Despite most CO_2-LSWAG studies indicating improvement in oil recovery in Table 7.1, some studies (Jiang et al., 2010; Ramanathan et al., 2015) with core samples came out with negative or neutral outcomes due to initial wettability being strongly water-wet which is not favourable for effective low salinity water injection. Furthermore, references can be made to the reviewed work by Ma and James (2022) based on CO_2-LSWAG laboratory experiments with no machine learning models reported.

Table 7.1. CO_2 LSWAG based on field, experimental, numerical simulation and machine learning models for oil recovery factor predictions.

Reference	Type of Porous Media	Type of Injection Fluid	Injection Scheme	Ultimate Oil Recovery, % OOIP	Experiment/ Field/ Numerical Simulation	Machine Learning Model
Al Quraisha et al. (2017)	Berea and Bentheimer Sandstone	$LSW + CO_2$ $HSW + CO_2$	LSW + HSW in secondary mode, CO_2 WAG in tertiary mode	82.40%	Experiment	No ML model
Al-Shalabi et al. (2014)	Carbonate Reservoir	$LSW1 + CO_2$ Injection	Tertiary mode	First Coreflood = 84.97% Second Coreflood = 93.65%	PHREEQC Simulation and Experiment	No ML model
Dang et al. (2014)	Sandstone	$CO_2 + LSW$	Secondary to tertiary modes	Incremental Oil Recovery of 4.5 – 9% of OOIP compared to CO_2 HSWAG	Field, Experiment and Numerical Simulation	No ML model
Naderi and Simjoo (2018)	Sandstone	$CO_2 + LSW$	Secondary to tertiary modes	16% of the OOIP on top of LSW and 30% of the OOIP on top of HSW	Compositional numerical simulation	No ML model
Zolfghari et al. (2013)	Sandstone	$CO_2 + LSW$	Tertiary Mode	92%	Experiment	No ML model
Teklu et al. (2016)	Low Permeability Carbonate Medium Permeability Berea Sandstone Ultra-Low Permeability Three Forks mudstone Core Disc	$LSW + CO_2$	Secondary to tertiary modes	Coreflood experiment 1 = 81.9% Coreflood experiment 2 = 83.4% Coreflood experiment 3 = 81.5%	Experiment	No ML model
Saxena (2017)	Sandstone	$LSW1 + CO_2$	Tertiary mode	39% OOIP	Compositional numerical simulation and Field	No ML model

ML = Machine Learning, LSW = Low Salinity Water, OOIP = Original Oil In Place

Al-Jifri et al. (2021) develop two new empirical equations for predicting oil recovery factor in waterflooded heterogeneous reservoirs based on these parameters of water injection rate, permeability anisotropy, water viscosity, and reservoir heterogeneity with no proxy correlations currently existing for CO_2-LSWAG which this study strive to achieve. Furthermore, Roustazadeh et al. (2022) developed three regression-based models including the support vector machine (SVM), extreme gradient boosting (XGBoost), and stepwise multiple linear regression (MLR) and various combinations of three databases to construct machine learning (ML) models and estimate the oil and/or gas recovery factor (RF). The following authors (Aliyuda et al., 2020; Alpak et al., 2019; Chen et al., 2020; Ibrahim et al., 2022; Sharma et al., 2010; Tahmasebi et al., 2020) also applied ML algorithms in predicting hydrocarbon recovery factors from different reservoirs but not with regards to the present study with CO_2 LSWAG.

In this study, a carbonate field model based on CO_2-LSWAG flooding was simulated using a compositional simulator with geochemical models incorporated, and then Multivariate Adaptive Regression Splines (MARS) and Group Method of Data Handling (GMDH) machine learning methods were used to develop proxy models for prediction of oil recovery factor. Therefore, this study advocates for the injection of low-salinity water alternating CO_2 as an EOR technique due to its high recovery factor for improving microscopic and macroscopic displacement efficiencies. The use of machine learning proxy models as prediction tools will also enhance the efficient full-field implementation of this technique in reducing the computational time associated with numerical simulations (Amar et al., 2021; Kalam et al., 2021) in carbonate reservoirs.

The structure of this paper is organised as follows. Section 2 is the methodology that describes the use of a compositional simulator coupled with fluid flow and geochemical modelling, as well as machine learning tools. Section 3 is the results and discussion that analyses the currently proposed method and its comparison to the use of a conventional simulator to optimise operational conditions. Section 4 concludes the study and spells out the major findings drawn from the present study.

2. Methodology

2.1 Modelling of CO_2-LSWAG

This paper's numerical method used a compositional simulator to generate various scenarios of the CO_2-LSWAG in a carbonate reservoir with SO_4

and carboxylate ion exchange. The assumptions made in modelling the CO_2-LSWAG in CMG compositional simulator under miscible conditions are:

1. Carboxylate ion exchange was solely responsible for the wettability alteration.
2. Multiphase multicomponent flow equations and EOS flash calculations were fully coupled to geochemical reactions.
3. Relative permeability sets of oil and water are altered by a scaled carboxylate ion exchange equivalent fraction.
4. Carboxylate ion exchange is also dependent on mineral dissolution and precipitation reactions.

2.2 Geochemical Reactions of CO_2-LSWAG

Carbonates with SO_4^{2-} and carboxylate ion exchange interpolation was carried out on sets of relative permeabilities. Equations (7.2) and (7.3) show the partitioning of CO_2 between oil and water in the reservoir (Dang et al., 2014).

The aqueous reactions considered in CO_2 LSWAG modelling are:

$$CO_{2(g)} \rightarrow CO_{2(aq)} \qquad (7.1)$$

$$CO_{2(aq)} + H_2O \leftrightarrow H^+ + HCO_3^- \qquad (7.2)$$

$$H^+ + OH^- \leftrightarrow H_2O \qquad (7.3)$$

$$CaCH_3COO^+ \leftrightarrow CH_3COO^- + Ca^{2+} \qquad (7.4)$$

$$CaSO_4 \leftrightarrow Ca^{2+} + SO_4^{2-} \qquad (7.5)$$

$$MgSO_4 \leftrightarrow Mg^{2+} + SO_4^{2-} \qquad (7.6)$$

The mineral reaction considered in CO_2-LSWAG modelling for calcite and dolomite are:

$$CaCO_3 + H^+ \leftrightarrow Ca^{2+} + HCO_3^- \qquad (7.7)$$

$$CaMg(CO_3)_2 + 2H^+ \leftrightarrow Ca^{2+} + 2HCO_3^- + Mg^{2+} \qquad (7.8)$$

The ion exchange considered in CO_2-LSWAG modelling is given as:

$$SO_4^{2-} + 2CH_3COO-X \leftrightarrow 2CH_3COO^- + SO_4 - X_2 \qquad (7.9)$$

2.3 Machine Learning Methods

The machine learning techniques of MARS and GMDH were used in this paper to predict Oil Recovery Factor (ORF) by varying the low salinity ionic

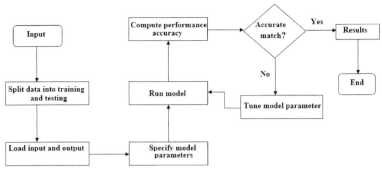

Fig. 7.1. Machine learning workflow.

concentrations of the water injected for the CO$_2$ LSWAG. Figure 7.1 shows the general workflow used for all the supervised machine-learning techniques. The datasets used were from the simulation results, which were then divided into training and testing data. The general workflow for all the supervised algorithms used in this study for CO$_2$-LSWAG ORF prediction as a single objective problem is shown in Fig. 7.1.

2.3.1 Multivariate Adaptive Regression Splines (MARS)

The multivariate nonparametric regression technique known as MARS was introduced by Friedman in 1991. This method utilises piecewise linear segments, or splines, to capture complex nonlinear relationships between input and target variables. The mathematical relationship between the basis function terms and the dependent variable in MARS is expressed in Eq. (7.10):

$$ORF = C_O + \sum_{k=1}^{M} C_k \beta_k(X), \qquad (7.10)$$

where *ORF* is the oil recovery factor as the target variable. There are several parameters that must be taken into account while simulating the interaction between two or more variables in a reservoir system. These parameters include the intercept (C_o), the number of basis function terms (M), and the vector of the kth basis function's unknown coefficients (C_k), where k is a value between 1 to *M*. The basis functions themselves are denoted by β_k and X is used to represent the input variables for the reservoir parameters.

According to Friedman's (1991) findings, the MARS technique uses multivariate spline functions as basis functions, which are represented in Eq. (7.11):

$$\beta_k(X) = \prod_{z=1}^{Z_n} \left[p_{zn} \cdot (X_{m(z,n)} - s_{zn}) \right]_+, \qquad (7.11)$$

In this context, we use the variable Z_n to represent the number of splits that resulted in the value of β_k. The variable p_{zn} is utilized as a sign indicator that can take on values ±1. The notation [.]+ is used to signify taking only the positive part. The function $m(z, n)$ labels the predictor variable, while s_{zn} represents the corresponding knot value of variable $X_{m(z,n)}$.

The MARS algorithm is a popular approach for developing models that can capture nonlinear relationships between variables. The algorithm typically consists of two steps, the forward and backward passes. In the forward pass, the algorithm investigates potential knots and begins with a model consists of just the constant term. This means that the initial model assumes that there is no relationship between the predictor variables and the response output variable. As the algorithm progresses, it gradually adds basis functions (BFs) to the model, which allows it to capture the nonlinearity between the predictor variables and the response output variable. The backward pass is carried out after the forward pass and is used to eliminate estimators that are irrelevant to the output variable. This is done to streamline the developed model eliminating the least significant BFs according to the generalised cross-validation (*GCV*) criterion given in Eq. (7.12). The *GCV* criterion is a widely used method for selecting the best subset of predictors that minimise the prediction error of the model. By removing the non-influential predictors, the model becomes more interpretable and easier to understand, while still accurately capturing the nonlinear relationships between the predictor variables and the response output variable.

$$GCV(\varphi) = \frac{\frac{1}{N}\sum_{i=1}^{N}(O_i - P_i)^2}{\left[1 - \frac{M(\varphi)}{N}\right]^2}, \qquad (7.12)$$

To achieve a balance between the model size and its fit to the dataset, a tuning parameter denoted as φ is utilised. The overall field measurements training dataset is represented by N, while O_i and P_i denote the observed and predicted values of the dataset, respectively. By adjusting the value of the tuning parameter, one can regulate the trade-off between the complexity of the model and its accuracy in capturing the dataset's patterns. $M(\varphi)$ is defined in Eq. (7.13) as the effective number of parameters employed in the model:

$$M(\varphi) = (\varphi + 1) + d.\varphi, \qquad (7.13)$$

The parameter d serves as a penalty or smoothing factor in the non-parametric MARS algorithm, where its value determines the number of basis functions and the smoothness of the estimated functions. A higher value of d leads to fewer basis functions and smoother estimates, whereas a lower value results in a larger model with more basis functions. For further information on the selection of d values and a comprehensive explanation of the non-parametric MARS algorithm, please refer to Friedman's work from 1991.

The format for the MARS equations expressed in terms of max(.) for univariate linear regression of the pth piece of variable x_k as the basis function is presented in Eq. (7.14). This mathematical expression demonstrates how $BF_p()$ is represented within the MARS framework. To ensure clarity, appropriate synonyms have been used, and sentences have been rearranged without altering the intended meaning.

$$BF_p(x_k) = \max(0, x_k - a_p) \text{ or } BF_p(x_k) = \max(0, a_p - x_k). \qquad (7.14)$$

In mathematical notation, the term max(.) indicates that solely the positive portion of the input is preserved while assigning a zero value to the negative portion, as presented in equation (7.15). Piecewise linear functions can be represented in the form of $\max(x_k - a_p)$, with the knot point specified at a particular value a_p.

$$\max(0, x_k - a_p) = \begin{cases} x_k - a_p, & x_k \geq a_p \\ 0, & \text{otherwise} \end{cases}, \qquad (7.15)$$

To create a local linear regression with continuous knots, the max (.) function is utilized. Through recursive spline fitting and splitting, the knots are best chosen. The Eq. (7.14) $BF_p(x_k)$ will only be zero when the second term in the equation exceeds zero, which is a crucial point to remember. Additionally, when modeling two variable interactions, the basis function $BF_{ps}()$ can be expressed using two univariate basis functions for x_k and x_i, as shown in Eq. (7.16):

$$BF_{ps}(x_k, x_i) = BF_p(x_k) \times BF_s(x_i). \qquad (7.16)$$

2.3.2 Group Method of Data Handling (GMDH)

In this study, GMDH is used to develop an explicit mathematical model with degrees of the polynomials to predict the oil recovery factor for CO_2-LSWAG under the miscible condition in a carbonate reservoir. Ivakhnenko (1966) cyberneticist proposed a GMDH polynomial neural network (PNN)) algorithm for constructing high-order regression-type polynomials. However, due to the self-organising nature of the GMDH algorithm (selecting its own input parameters automatically based on the parameter's contribution to the final output), the GMDH modelling approach is able to overcome many artificial neural network (ANN) limitations as an alternative method (Lv et al., 2023). GMDH employs optimisation criteria to determine network connectivity, network size, and the coefficient for the optimum model with model reduction with less human intervention (Ivakhnenko, 1971). The input and output relationship are expressed in polynomial form, with the model automatically selecting the most influential parameters (Farlow, 1984). In this study, GMDH is used to develop a mathematical model for predicting the CO_2-LSWAG oil recovery factor. Ivakhnenko (1971) applied the Kolmogorov-Gabor polynomials theory function to find the output parameter, which is expressed in Eq. (7.17):

$$ORF^{GMDH} = a_o + \sum_{i=1}^{n} a_i x_i + \sum_{i=1}^{n}\sum_{j=1}^{n} a_i a_j x_i x_j + \sum_{i=1}^{n}\sum_{j=1}^{n}\sum_{k=1}^{n} a_i a_j a_k x_i x_j x_k + \cdots \quad (7.17)$$

where, $a_{ij\ldots k}$, $x_{ij\ldots k}$, ORF, and n represent polynomial coefficients, input parameters, the predicted values and the number of variables, respectively.

From Eq. (7.17) the polynomial variables are obtained by employing least square minimisation analysis as expressed in Eq. (7.18):

$$\delta_j^2 = \sum_{i=1}^{N} (ORF_i^{GMDH} - y_i)^2 \quad (7.18)$$

where N, y_i, and ORF_i^{GMDH} represent the number of training data points, actual values, and predicted values, respectively.

2.3.3 Performance Metrics

The accuracy of the prediction from the machine learning methods was measured using the coefficient of determination (R^2), mean squared error (MSE) and Pearson's correlation coefficient (r) to understand the target against variations of individual independent variables.

The R^2 is a statistical index with the best model prediction having an optimum value closer to 1, MSE gets closer to zero for an accurate model, and r can be -1 or $+1$, indicating a parametric effect on output (Brantson et al., 2019). Equations (7.19), (7.20), and (7.21) illustrate the mathematical expressions used for R^2, MSE and r, respectively:

$$R^2 = \left(\frac{\sum_{i=1}^{n} (ORF_{observed} - \overline{ORF}_{observed\ mean}) \times (ORF_{predicted} - \overline{ORF}_{predicted\ mean})}{\sqrt{\sum_{i=1}^{n} (ORF_{observed} - \overline{ORF}_{observed\ mean})^2 \times (ORF_{predicted} - \overline{ORF}_{predicted\ mean})^2}} \right)^2 \quad (7.19)$$

$$MSE = \frac{1}{n}\sum_{i=1}^{n} (ORF_{measured} - ORF_{predicted})^2 \quad (7.20)$$

$$r = \frac{\sum_{i=1}^{n}\left(ORF_{observed} - \overline{ORF}_{observed\ mean}\right) \times \left(ORF_{predicted} - \overline{ORF}_{predicted\ mean}\right)}{\sqrt{\sum_{i=1}^{n}\left(ORF_{observed} - \overline{ORF}_{observed\ mean}\right)^2 \times \sum_{i=1}^{n}\left(ORF_{predicted} - \overline{ORF}_{predicted\ mean}\right)^2}} \quad (7.21)$$

where, n = total number of oil recovery factor data points observed, $ORF_{observed}$ = oil recovery factor observed data points, $\overline{ORF}_{observed\ mean}$ = mean of the oil recovery factor observed data points, and $\overline{ORF}_{predicted\ mean}$ = mean of the oil recovery factor predicted data points.

2.3.4 Dataset Standardisation

The CO_2-LSWAG modeling process involves the use of data with varying physical units. These physical units could include meters, grams, seconds, or any other units that are relevant to the data being used. In order to ensure that the data is comparable and can be used to generate accurate models, normalisation is applied. Normalisation is a technique that scales the data to a common range of values. In this particular study, both the input and output vectors were normalised in the interval [−1, 1]. This range was chosen because it is a common range of values used in normalisation techniques. The normalisation process was implemented to ensure that there is a constant variation in the data as expressed in Eq. (7.22). By normalising the data, the range of values for both input and output vectors is restricted to a specific range, thus making it easier to compare and analyse the data. It is important to note that normalisation is not always necessary, but it is often used when working with data that has varying physical units. By normalising the data, it is possible to remove any biases that may arise due to differences in physical units. This can help to ensure that the models generated from the data are accurate and reliable.

$$y_i = y_{min} + \frac{(y_{max} - y_{min}) \times (x_i - x_{min})}{(x_{max} - x_{min})}, \quad (7.22)$$

The resulting normalised data is denoted by y_i, where the maximum and minimum values (y_{max} and y_{min}) of the normalised data are set at 1 and −1, respectively, the measured values (x_i) are scaled to a range of −1 to 1, with x_{min} and x_{max} the measured data serving as the lower and upper bounds of this range. This method of normalisation is commonly used.

3. Results and Discussion

3.1 Numerical Model Description

This study employed an inverted five-spot injection carbonate model to simulate CO_2-LSWAG oil recovery at the field scale. The reservoir had dimensions of approximately 7400 × 7400 × 28 ft³ along the Y, X, and Z directions. A total of 12,150 grid blocks were utilised, comprising of 45 blocks in both X and Y directions and six blocks in the Z direction. The simulation utilised flexible grids and precise corner point geometry to accurately represent the geological features and description of the reservoir, particularly for heterogeneities and wells. This approach was preferred over conventional cartesian grids. The reservoir exhibited heterogeneity with variations in porosity and permeability. Porosity values ranged between 9.5%–29%, while permeability differed in all directions with the vertical permeability lower than the horizontal permeability. The active fluid phases used in the model were water and oil, which are commonly found in reservoirs. The porosity and permeability variations depicted in Figs. 7.2a and 7.2b provide valuable insights into the characteristics of the reservoir model. Additionally, Fig. 7.3 depicts the oil and water relative permeability curves used for the CO_2-LSWAG simulation for high and low salinity as a function of water saturation. The aqueous or mineral reaction structure selected was for

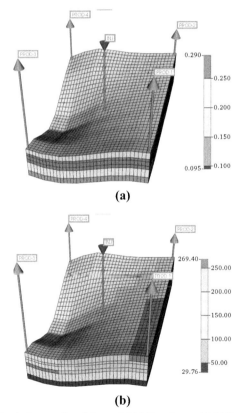

Fig. 7.2. Petrophysical properties of the model (a) Porosity model (b) Permeability model.

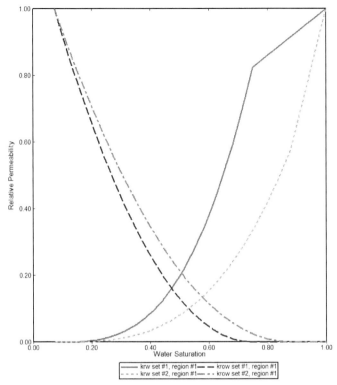

Fig. 7.3. Low and high salinity water and oil relative permeability curves (set #1 is for the original relative permeability plot while set #2 is for the LSW relative permeability plot).

carbonates with SO_4^{2-} simple relative permeability interpolation. The component wizard used linear interpolation for SO_4^{2-} concentration to effect the change.

Upon completion of the reservoir simulation model, well completion procedures were implemented. In this study, we determined that the inverse five-spot pattern was the most optimal approach for the reservoir simulation. To achieve maximum effectiveness, it was ensured that all six grid layers were perforated for both production and injection wells, with vertical wells for efficient oil extraction. The placement of the injection well in the central location allowed for efficient fluid sweep throughout the reservoir, moving the oil towards the production wells situated at the corners, resulting in a high areal sweep efficiency. The sweeping motion of the injected fluid played a significant role in improving oil recovery in the reservoir. The implementation of the inverse five-spot pattern and vertical wells in all six grid layers of the model enabled efficient extraction of the oil.

3.2 Input and Target Dataset

A simulation study was carried out to obtain a simulation dataset for input and output for training and testing to establish a mapping relationship. Table 7.2 shows the reservoir parameters that were used in running the carbonate simulation model. Additionally, Table 7.3 shows the initial run whose result was used as the base case for the sensitivity run for varied low salinity water ranges between 200–3000 ppm. The total dataset obtained within the low salinity water ranges was 1385 sample points, with 1,107 used for training and the rest used for testing the model. Furthermore, Fig. 7.4a shows the average pressure for the base case and CO_2-LSWAG scenarios.

Table 7.2. Reservoir parameters for the carbonate model.

Parameter	Value
Reference pressure, psi	3337
Reference depth, ft	8596
Water-Oil contact depth, ft	8950
Permeability, mD	29.76–269.40
Porosity, %	9.5–29.
Top of Reservoir Sand, ft	8596
Swcon	0.076
Soirw	0.1434
1-Sorw	0.75
1-Soirw	0.8566
pH	5.22
Water Injection Rate, bbl/day	7,000
Gas Injection Rate, MMft³/day	30
Injector Bottomhole Pressure, psi	6,710
Formation Water Salinity, ppm	90,044.10
Equation of State (EOS)	Peng Robinson

Table 7.3. Base case for the formation and injected water.

Species	Formation Water, ppm	Base Case Injected Water, ppm
Ca^{2+}	18492	26
Mg^{2+}	2320	77
Na^+	68520.1	660
HCO_3^-	100	0
SO_4^{2-}	612	46.8

Recovery factor prediction can be a function of multiple variables of the reservoir characteristics to be determined. The operational controllable input parameters for the MARS and GMDH models are average reservoir pressure (ARP), cumulative gas injection (CGI), cumulative low salinity water injection (CLSWI), gas cumulative production (GCP), oil cumulative production (OCP), gas injection rate (GIR), low salinity water injection rate (LSWIR), and the target is the ORF for CO_2-LSWAG as shown in Table 7.4. The LSW and CO_2 injection rate employed in the carbonate model for the simulation period is shown in Fig. 7.4b.

For constructing the proxy machine learning models, the oil saturation maps obtained for the 15 years of CO_2-LSWAG injection oil recovery are shown in Figs. 7.5a, 7.5b, and 7.5c for 5, 10, and 15 years, respectively. It was observed that CO_2 injection lowers the viscosity of the crude oil after it becomes miscible with it. Figure 7.5 shows the initial and final oil saturation of the reservoir, depicting how the CO_2-LSWAG injection lowered the viscosity to recover more oil. It can be seen from the final oil saturation plot in Fig. 7.5c of the CO_2-LSWAG that the injection scheme lowered the viscosity of the reservoir oil, enabling more oil to be recovered.

Table 7.4. Dataset ranges for CO_2-LSWAG oil recovery factor prediction.

Variable/Unit	Minimum	Average	Maximum	Standard Deviation	Data Type
Average reservoir pressure (psi)	3349.11	5187.38	6679.61	1298.05	Input
Cumulative gas injection (MMMSCF)	0.00	12.12	25.64	5.28	Input
Cumulative low salinity water injection (MMbbl)	0.00	6.83	18.48	4.24	Input
Gas cumulative production (MMMSCF)	0.00	7.61	17.02	3.16	Input
Oil cumulative production (MMbbl)	0.00	13.73	15.96	3.17	Input
Gas injection rate (MMSCFD)	0.00	13.81	30.00	4.77	Input
Low salinity water injection rate (Mbbl/day)	0.00	5.09	7.17	1.58	Input
Oil recovery factor (%)	0.00	72.38	84.07	16.68	Output

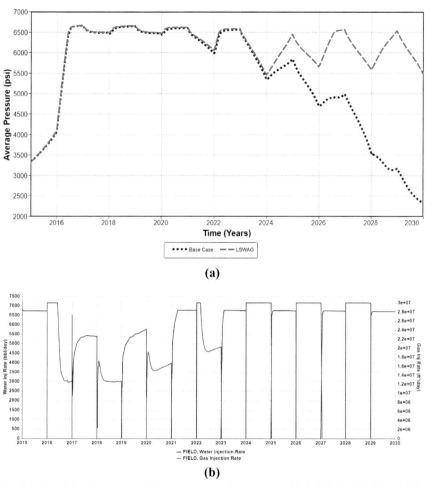

Fig. 7.4. Simulation Parameters: (a) Average pressure base case and CO_2 LSWAG, (b) LSW and CO_2 injection rate.

3.3 MARS Modelling

MARS modelling combines both recursive spline fitting and partitioning that maintains the positive component of the combination while having less influence on non-contributing properties (Brantson et al., 2018). As shown in Table 7.4, all the inputs and target will be loaded into the MARS algorithm for developing the trained model for spline fitting optimisation computations. Also, the MARS model automatically selects the most influential input parameters that affect the target response leaving out the non-contributing variables. Furthermore, the maximum basis functions, maximum interactions and degrees of freedom for knot optimisation used

144 *Data Science and Machine Learning Applications in Subsurface Engineering*

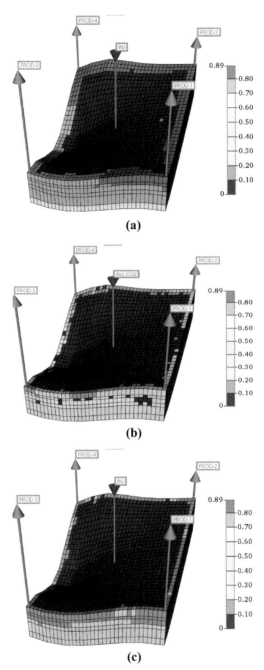

Fig. 7.5. Oil saturation maps for the simulation period: (a) oil saturation at 5 years, (b) oil saturation at 10 years, (c) oil saturation at 15 years.

Table 7.5. Basis functions and corresponding equations of MARS model.

Basis Functions (BF)	Equations
BF1	max (0, OCP − 1.03286 × 10^7)
BF2	max (0, 1.03286 × 10^7 − OCP)
BF4	max (0, 4.09049 × 10^8 − GCP)
BF5	max (0, GCP − 5.33908 × 10^9)
BF7	max (0, GCP − 4.43216 × 10^9)
BF10	max (0, 3802.39 − ARP)
BF12	max (0, 3.97977 × 10^9 − CGI)
BF13	max (0, OCP − 2.79576 × 10^6)
BF15	max (0, LSWIR − 7042.81)
BF17	max (0, OCP − 7.16527 × 10^6)
BF19	max (0, OCP − 1.52833 × 10^7)

are 20, 1 and 3, respectively. Hence the relative importance of the five input variables used are oil cumulative production (100%), gas cumulative production (2.22%), cumulative gas injection (2.36%), the water injection rate (1.63%), and average pressure (1.46%). The relevance factor computed for all the parameters chosen automatically has a positive impact on the ORF as the target. The ideal developed MARS model for CO$_2$-LSWAG is expressed in Eq. (7.23). Also, Table 7.5 indicates the basis functions and corresponding equations of the ORF MARS model written in Eq. (7.23):

$$\begin{aligned}ORF = {} & 25.7653 + 1.84934 \times 10^{-6} \times BF1 - 1.57532 \times 10^{-6} \times BF2 + 1.00632 \times 10^{-8} \times BF4 \\ & - 2.33007 \times 10^{-9} \times BF5 + 2.36461 \times 10^{-9} \times BF7 + 0.00204046 \times BF10 \\ & - 3.65011 \times 10^{-9} \times BF12 + 3.67347 \times 10^{-6} \times BF13 + 0.0775678 \times BF15 \\ & - 3.02324 \times 10^{-7} \times BF17 - 4.68672 \times 10^{-7} \times BF19\end{aligned} \quad (7.23)$$

Figures 7.6a to 7.6d show the cross plots for the MARS training and testing models used in this study. The results show both the training and testing data points are close to the ideal line, which is an indication of the MARS model's robustness. Three independent datasets (low salinities of 573.86 ppm, 1250.51 ppm, and 2949.15 ppm) were used to test the MARS model to verify its accuracy and reliability. Furthermore, the CO$_2$-LSWAG recovery factor prediction can also be assessed from Table 7.6, showing the statistical performance. The boxplot in Fig. 7.6e shows the residuals for both MARS training and testing sets. It can be observed that the residuals do not vary significantly from zero. It can be stated that the training model established can predict the testing datasets within acceptable accuracy predictions. It is shown that the mean and median lie close to zero for a good model. Figure 7.6e also shows the MARS model residual distributions and outliers.

146 *Data Science and Machine Learning Applications in Subsurface Engineering*

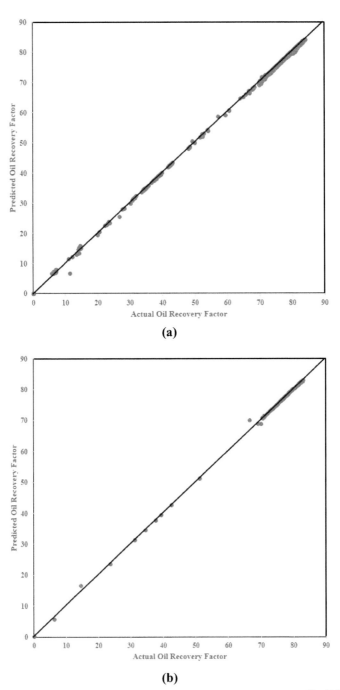

Fig. 7.6 contd. ...

...Fig. 7.6 contd.

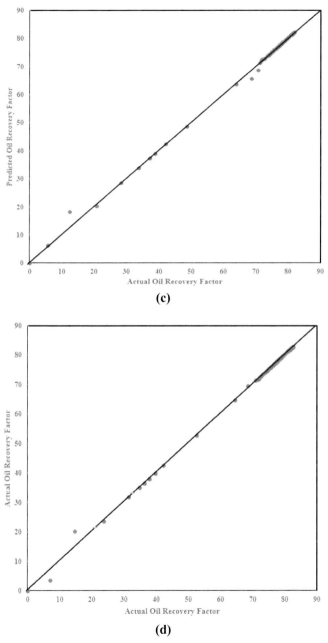

(c)

(d)

Fig. 7.6 contd. ...

...Fig. 7.6 contd.

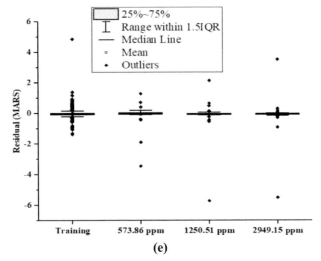

Fig. 7.6. MARS training and testing models: (a) Crossplot for the MARS training model, (b) Crossplot for the MARS testing for 573.86 ppm, (c) Crossplot for the MARS testing for 1250.51 ppm, (d) Crossplot for the MARS testing for 2949.15 ppm, (e) Boxplot residuals for MARS training and testing.

Table 7.6. Train and test results for MARS model.

Models	Training	Testing (573.86 ppm)	Testing (1250.51 ppm)	Testing (2949.15 ppm)
MSE	0.0564	0.2023	0.5541	0.4518
R^2	0.9998	0.9996	09993	0.9995

3.4 GMDH Modelling

In building the GMDH model, the most influential input parameters are usually automatically selected that have an influence on the target variable. Three input variables that have the most influence on the target are CLSWI, OCP, and LSWIR. The number of layers and neurons used are 2 and 2, respectively, for GMDH model computations. The final architecture obtained was after a series of optimisation processes by observing the network performance.

Figures 7.7a to 7.7d show the cross plots for both the GMDH training and testing models used in this study. It can be seen that both the training and testing data points are close to the ideal line, which is an indication of the GMDH model's robustness. Furthermore, the CO_2-LSWAG recovery factor prediction can also be assessed from Table 7.7, showing the statistical

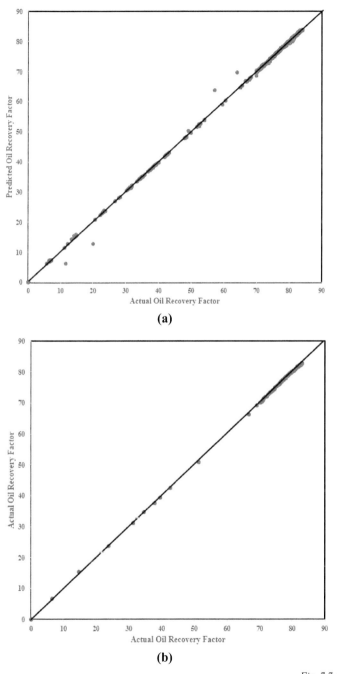

Fig. 7.7 contd. ...

...*Fig. 7.7 contd.*

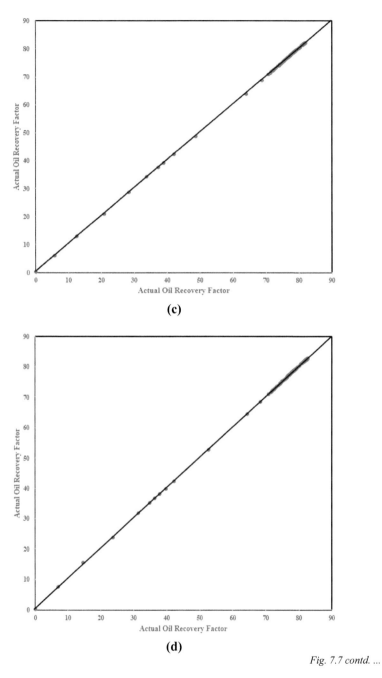

Fig. 7.7 contd. ...

...Fig. 7.7 contd.

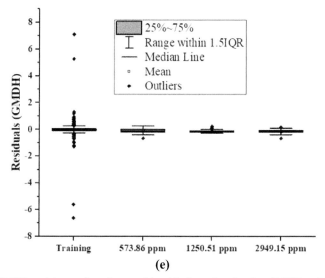

Fig. 7.7. GMDH training and testing models: (a) Crossplot for the GMDH training model, (b) Crossplot for the GMDH testing for 573.86 ppm, (c) Crossplot for the GMDH testing for 1250.51 ppm, (d) Crossplot for the GMDH testing for 2949.15 ppm, (e) Boxplot residuals for GMDH training and testing.

Table 7.7. Training and testing results for the GMDH model.

Models	Training	Testing (573.86 ppm)	Testing (1250.51 ppm)	Testing (2949.15 ppm)
MSE	0.1674	0.0239	0.0219	0.0231
R^2	0.9999	0.9999	0.9999	0.9999

performance. Also, it can be observed from Table 7.7 that the testing data performed better than the MARS model testing. The boxplot in Fig. 7.7e shows the residuals for both GMDH training and testing sets. It can be observed that the residuals do not vary significantly from zero. It can be stated that the training model established can predict the testing datasets within acceptable accuracy predictions. It can also be seen that the mean and median lie close to zero for a good model. Figure 7.7e also shows the residual distributions and least outliers for GMDH testing.

Figure 7.8 shows the GMDH topology used to develop the *ORF* equations. The ORF was formed from three variables from the input layer. The OCP variable combines with CLSWI and LSWIR in the input layer to form two variables in the hidden layer before combining them to build the target layer.

The summary of the GMDH model's equations for two layers is expressed in Eqs. (7.24) to (7.26) as:

Layer #1
Number of neurons: 2

$$A = a_0 + a_1 \times OCP + a_2 \times CLSWI + a_3 \times OCP \times CLSWI + a_4 \times (OCP)^2 + a_5 \times (CLSWI)^2 \quad (7.24)$$

$a_0 = 0.03088$ $a_1 = 0.94845$ $a_2 = 0.04723$
$a_3 = 0.05133$ $a_4 = 0.02253$ $a_5 = 0.00212$

$$B = -b_0 + b_1 \times LSWI + b_2 \times OCP - b_3 \times OCP \times LSWIR - b_4 \times (LSWIR)^2 + b_5 \times (OCP)^2 \quad (7.25)$$

$b_0 = 0.00396$ $b_1 = 0.00360$ $b_2 = 0.99753$
$b_3 = 0.00431$ $b_4 = 0.00041$ $b_5 = 0.01058$

Layer #2
Number of neurons: 1

$$ORF = -c_0 + c_1 \times B + c_2 \times A + c_3 \times A \times B - c_4 \times (B)^2 + c_5 \times (OCP)^2 \quad (7.26)$$

$c_0 = 0.00038$ $c_1 = 0.61360$ $c_2 = 0.37942$
$c_3 = 409.55682$ $c_4 = 205.30898$ $c_5 = 204.23896$

where A and B are virtual independent inputs or nodal variables used in the GMDH neural network.

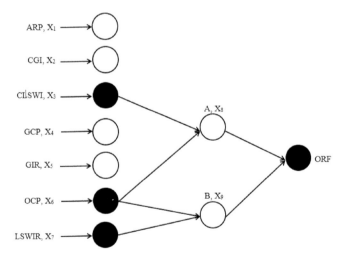

Fig. 7.8. Schematic for the GMDH topology.

3.5 Numerical Simulator and Machine Learning Computational Time

Figure 7.9 shows the simulation time for the CMG commercial numerical simulator and the machine learning algorithms using Intel core i5 – 11400H, 4.5 GHz 6Core, and 16 GB of RAM. It can be seen that the numerical simulation took a long runtime to execute. Also, this significant reduction in simulation time indicates the robustness of the machine-learning proxy models (MARS and GMDH). Furthermore, the proxy models can serve as substitutes for direct simulation of CO_2-LSWAG for design parameters within the ranges considered. In summary, using a GPU (graphics processing unit) could further improve the computational time of this machine-learning algorithm. In comparison to using a GPU, this method may offer even greater benefits in terms of speed and efficiency when running multiple simulations at the same time.

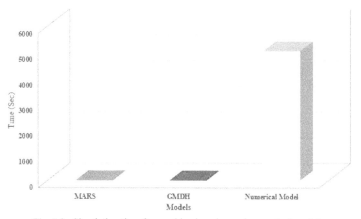

Fig. 7.9. Simulation time for machine learning and numerical models.

4. Conclusion

The problem addressed in this study is the lack of attention and research on using machine learning techniques to use CO_2-LSWAG for EOR in carbonate reservoirs. The quality and composition of the water used in most EOR processes are not adequately considered, and there is a need for fast computational methods for predicting future recovery factors. We used multiphase multicomponent flow equations, geochemical modelling, and compositional simulation datasets to build proxy models for predicting oil recovery factors in carbonate reservoirs for CO_2-LSWAG EOR. Two supervised machine learning models (MARS and GMDH) were used to analyse the data generated by these simulations under various low-salinity conditions. The conclusion from the study are as follows:

1. The study showed that implementing CO_2-LSWAG at a field scale can significantly increase the recovery of oil by mainly reducing viscous fingering.
2. The GMDH model gave better testing predictions as compared to the MARS prediction model with derived mathematical correlations for the CO_2 LSWAG oil recovery factor.
3. The machine learning models predicted the outputs from the commercial simulator by significantly reducing simulation time and generalising without compromising the required accuracy.
4. It is recommended that CO_2-LSWAG should be implemented on a field scale based on the current study to increase recovery while reducing operational costs.

Acknowledgment

We express our appreciation to the anonymous reviewers whose valuable inputs have significantly contributed to the success of this research. Additionally, our gratitude goes to the Computer Modelling Group (CMG) for providing us with the commercial software that was instrumental to the completion of this project. Lastly, we acknowledge the indispensable support of the University of Mines and Technology, Tarkwa, Ghana, GNPC School of Petroleum Studies, and the Petroleum and Natural Gas Engineering Department.

References

Afzali, S., Rezaei, N. and Zendehboudi, S.A. 2018. Comprehensive review on enhanced oil recovery by water alternating gas (WAG) injection. *Fuel* 227: 218–246. https://doi.org/10.1016/j.fuel.2018.04.015.

Ahmed, T. 2018. *Reservoir Engineering Handbook* (4th Edn.). Elsevier Inc.

Al-Abri, H., Pourafshary, P., Mosavat, N. and Al Hadrami, H. 2019. A study of the performance of LSWA CO_2 EOR technique on improvement of oil recovery in sandstones. *Petroleum* 5(1): 58–66. https://doi.org/10.1016/j.petlm.2018.07.003.

Alaidan, A. and Mamore, D. 2010. SWACO_2 and WACO_2 efficiency improvement in carbonate cores by lowering water salinity. *In*: *Canadian Unconventional Resources and International Petroleum Conference*. https://doi.org/10.2118/137548-MS.

Aleidan, A. and Mamora, D. 2011. Miscible CO_2 injection in highly heterogeneous carbonate cores: experimental and numerical simulation studies. *In*: *SPE Middle East Oil and Gas Show and Conference*. https://doi.org/10.2118/141469-MS.

Aliyuda, K., Howell, J. and Humphrey, E. 2020. Impact of geological variables in controlling oil-reservoir performance: an insight from a machine-learning technique. *SPE Reservoir Evaluation & Engineering* 23(04): 1314–1327. https://doi.org/10.2118/201196-PA.

Al-Jifri, M., Al-Attar, H. and Boukadi, F. 2021. New proxy models for predicting oil recovery factor in waterflooded heterogeneous reservoirs. *Journal of Petroleum Exploration and Production* 11(3): 1443–1459. https://doi.org/10.1007/s13202-021-01095-4.

Alpak, F.O, Araya–Polo, M. and Onyeagoro, K. 2019. Simplified dynamic modeling of faulted turbidite reservoirs: a deep-learning approach to recovery-factor forecasting for exploration. *In*: *SPE Reservoir Evaluation & Engineering* 22(04): 1240–55. https://doi.org/10.2118/197053-PA.

AlQuraishi, A.A. and Shokir, E.M. El-M. 2011. Experimental investigation of miscible CO_2 flooding, *Journal of Petroleum Science and Technology* 29(19). https://doi.org/10.1080/10916461003662976.

AlQuraishi, A.A., Amao, A.M., Al-Zahrani, N.I., AlQarni, M.T. and AlShamrani, S.A. 2019. Low salinity water and CO_2 miscible flooding in berea and bentheimer sandstones. *Journal of King Saud University – Engineering Sciences* 31(3): 286–295. https://doi.org/10.1016/j.jksues.2017.04.001.

Al-Shalabi, E.W., Sepehrnoori, K. and Pope, G. 2014. Geochemical investigation of the combined effect of injecting low salinity water and carbon dioxide on carbonate reservoirs. *Energy Procedia* 63: 7663–7676. https://doi.org/10.1016/j.egypro.2014.11.800.

Al-Shalabi, E., Sepehrnoori, K. and Pope, G. 2016. Numerical modelling of combined low salinity water and carbon dioxide in carbonate cores. *Journal of Petroleum Science and Engineering* 137: 157–171. https://doi.org/10.1016/j.petrol.2015.11.021.

Amar, M.N., Ghahfarokhi, A.J., Ng, C.S.W. and Zeraibi, N. 2021. Optimization of WAG in real geological field using rigorous soft computing techniques and nature-inspired algorithms. *Journal of Petroleum Science and Engineering* 206(109038): 1–13. https://doi.org/10.1016/j.petrol.2021.109038.

Belazreg, L. and Mahmood, S.M. 2020. Water alternating gas incremental recovery factor prediction and WAG pilot lessons learned. *Journal of Petroleum Exploration and Production Technology* 10: 249–269. https://doi.org/10.1007/s13202-019-0694-x.

Brantson, E.T., Ju, B., Omisore, O.O., Wu, D., Aphu, E.S. and Liu, N. 2018a. Development of machine learning predictive models for history matching tight gas carbonate reservoir production profiles. *Journal of Geophysics and Engineering* 15(5): 7–12. https://doi.org/10.1088/1742-2140/aaca44.

Brantson, E.T., Ju, B., Omisore, B.O., Wu, D., Selase, A.E. and Liu, N. 2018b. Development of machine learning predictive models for history matching tight gas carbonate reservoir production profiles. *Journal of Geophysics and Engineering* 15(5): 2235–2251.

Brantson, E.T., Ju, B., Ziggah, Y.Y., Akwensi, P.H., Sun, Y., Wu, D. and Addo, B.J. 2019. Forecasting of horizontal gas well production decline in unconventional reservoirs using productivity, soft computing and swarm intelligence models. *Natural Resources Research* 28: 717–756.

Chen, Y., Zhu, Z., Lu, Y., Hu, C., Gao, F., Li, W., Sun, N. and Feng, T. 2020. Reservoir recovery estimation using data analytics and neural network based analogue study. *In*: *SPE/IATMI Asia Pacific Oil & Gas Conference and Exhibition*. https://doi.org/10.2118/196487-MS.

Christensen, J.R., Stenby, E.H. and Skauge, A. 1998. Review of WAG field experience. *In*: *Proceedings of the International Petroleum Conference and Exhibition of Mexico*. https://doi.org/10.2118/90589-MS.

Dang, C.T.Q, Nghiem, L.X, and Chen, Z. 2014. CO_2 low salinity water alternating gas: a new promising approach for enhanced oil recovery. *In*: *SPE Improved Oil Recovery Symposium*, 1–19. https://doi.org/10.2118/169071-MS.

Farlow, S.J. 1984. *Self-Organizing Methods in Modeling GMDH Type Algorithms*. New York: Marcel-Dekker. CRC Press.

Friedman, J.H. 1991. Estimating Functions of Mixed Ordinal and Categorical Variables using Adaptive Splines (pp. 1–42). Department of Statistics, Stanford Univ.

Gbadamosi, A.O., Radzuan, J., Manan, M.A., Agi, A. and Adeyinka, S.Y. 2019. An overview of chemical enhanced oil recovery: recent advances and prospects. *Int. Nano Lett.* 9: 3–10. https://doi.org/10.1007/s40089-019-0272-8.

Hamouda, A.A. and Pranoto, A. 2016. Synergy between low salinity water flooding and CO_2 for EOR in chalk reservoirs. *In: Proceedings of the SPE EOR Conference at Oil and Gas West Asia, 2016; Society of Petroleum Engineers*. TX, USA: Richardson.

Ibrahim, A.F., Alarifi, S.A. and Elkatatny, S. 2022. Application of machine learning to predict estimated ultimate recovery for multistage hydraulically fractured wells in niobrara shale formation. *Computational Intelligence and Neuroscience* 1–10. http://dx.doi.org/10.1155/2022/7084514.

Ivakhnenko, A.G. 1966. Group method of data handling a rival of the method of stochastic approximation. *Soviet Automatic Control* 1(13): 43–71.

Ivakhnenko, A.G. 1971. Polynomial theory of complex system. *IEEE Transactions on System, Man and Cybernetics* 1(4): 364–378. 10.1109/TSMC.1971.4308320.

Jaber, A.K., Alhuraishawy, A.K. and AL-Bazzaz, W.H. 2019. A data-driven model for rapid evaluation of miscible CO_2-WAG flooding in heterogeneous clastic reservoirs. *In: Proceedings of the SPE Kuwait Oil & Gas Show and Conference*, 13–16. https://doi.org/10.2118/198013-MS.

Jiang, H., Nuryaningsih, L. and Adidharma, H. 2010. The effect of salinity of injection brine on water alternating gas performance in tertiary miscible carbon dioxide flooding: experimental study. *SPE Western Regional Meeting*. https://doi.org/10.2118/132369-MS.

Kalam, S., Khan, R.A., Khan, S., Faizan, M., Amin, M., Ajaib, R. and Abu-Khamsin, S.A. 2021. Data-driven modelling approach to predict the recovery performance of low-salinity waterfloods. *Natural Resources Research* 30: 1697–1717. https://doi.org/10.1007/s11053-020-09803-3.

Kharecha, P.A. and Hansen, J.E. 2008. Implications of peak oil for atmospheric CO_2 and climate. *Global Biogeochem* 22: 6–10. https://doi.org/10.48550/arXiv.0704.2782.

Kulkarni, M.M. and Rao, D.N. 2004. Experimental investigation of various methods of tertiary gas injection. *Paper Presented at the SPE Annual Technical Conference and Exhibition*. https://doi.org/10.2118/90589-MS.

Kumar, H., Shehata, A. and Nasr-El-Din, H. 2016. Effectiveness of low salinity and CO_2 flooding hybrid approaches in low permeability sandstone reservoirs. *In: SPE Trinidad and Tobago Section Energy Resources Conference*. https://doi.org/10.2118/180875-MS.

Lv, Q., Zhou, T., Zheng, R., Nakhaei-Kohani, R., Riazi, M., Hemmati-Sarapardeh, A., Li, J. and Wang, W. 2023. Application of group method of data handling and gene expression programming for predicting solubility of CO_2-N_2 gas mixture in brine. *Fuel* 332(6): 126025, 10.1016/j.fuel.2022.126025.

Ma, S. and James, L.A. 2022. Literature review of hybrid CO_2 low salinity water-alternating-gas injection and investigation on hysteresis effect. *Energies* 15(21): 7891. https://doi.org/10.3390/en15217891.

Mandadige, S.P., Ranjith, P.G., Tharaka, D.R., Ashani, S.R., Koay, A. and Choi, X. 2016. A review of CO_2-enhanced oil recovery with a simulated sensitivity analysis. *Energies* 9(7): 7–22. https://doi.org/10.3390/en9070481.

Naderi, S. and Simjoo, M. 2019. Numerical study of low salinity water alternating CO_2 injection for enhancing oil recovery in a sandstone reservoir: coupled geochemical and fluid flow modeling. *Journal of Petroleum Science and Engineering* 173: 279–286. https://doi.org/10.1016/j.petrol.2018.10.009.

Pourafshary, P. and Moradpour, N. 2019. Hybrid EOR methods utilizing low-salinity water. Enhanc. Oil Recovery Process. New Technol. 8: 25.

Ramanathan, R., Shehata, A. and Nasr-El-Din, H. 2015. Water alternating CO_2 injection process-does modifying the salinity of injected brine improve oil recovery? *In*: *Proceedings of the OTC Brasil, Rio de Janeiro, Brazil*, 27 October; *Offshore Technology Conference: Rio de Janeiro, Brazil*. https://doi.org/10.4043/26253-MS.

Ramanathan, R., Shehata, A. and Nasr-El-Din, H. 2016. Effect of rock aging in oil recovery during water-alternating-CO_2 injection process. *In*: *SPE Improved Oil Recovery Conference*. https://doi.org/10.2118/179674-MS.

Robertson, E.P. 2007. Low-Salinity Waterflooding to Improve Oil Recovery-Historical Field Evidence. *Paper Presented at the Annual Technical Conference and Exhibition*, 11–14, https://doi.org/10.2118/109965-MS.

Roustazadeh, A., Ghanbarian, B., Shadmand, M.B., Taslimitehrani, V. and Lake, L.W. 2022. *Estimating Oil and Gas Recovery Factors via Machine Learning: Database-Dependent Accuracy and Reliability*. Preprint. http://dx.doi.org/10.48550/arXiv.2210.12491.

Santos, R., Loh, W., Bannwart, A. and Trevisan, O. 2014. An overview of heavy oil properties and its recovery and transportation methods. *Brazilian Journal of Chemical Engineering* 31(3): 571–576. http://dx.doi.org/10.1590/0104-6632.20140313s00001853.

Saxena, K. 2017. *Low Salinity Water Alternate Gas Injection Process for Alaskan Viscous Oil EOR* [Master's Thesis, University of Alaska]. University of Alaska Fairbanks. http://hdl.handle.net/11122/7638.

Sharma, A., Srinivasan, S. and Lake, L.W. 2010. Classification of oil and gas reservoirs based on recovery factor: a data-mining approach. *In*: *SPE Annual Technology Conference & Exhibition* 1: 50–70. https://doi.org/10.2118/130257-MS.

Sheng, J.J. 2011. *Modern Chemical Enhanced Oil Recovery: Theory and Practice*. Elsevier Publishing Corporation.

Sheng, J.J. 2013. *Enhanced Oil Recovery Field Case Studies*. Waltham, Mass.: Gulf Professional Publishing.

Stalkup, F.I. 1984. *Miscible Displacement (SPE Monograph Series)*. Society of Petroleum Engineers.

Stosur, G.J. 2003. EOR: Past, Present, and What the Next 25 Years May Bring. *Paper Presented at the SPE International Improved Oil Recovery Conference in Asia Pacific*, Kuala Lumpur, Malaysia. https://doi.org/10.2118/84864-MS.

Sun, X., Zhang, Y., Chen, G. and Gai, Z. 2017. Application of nanoparticles in enhanced oil recovery: a critical review of recent progress. *Energies* 10(3): 1–6. https://doi.org/10.3390/en10030345.

Tahmasebi, P., Kamrava, S., Bai, T. and Sahimi, M. 2020. Machine learning in geo-and environmental sciences: From small to large scale. *Advances in Water Resources* 142(11): 103619. http://dx.doi.org/10.1016/j.advwatres.2020.103619.

Teklu, T.W., Alameri, W., Graves, R.M., Kazemi, H. and AlSumaiti, A.M. 2016. Low salinity water-surfactant-CO_2 EOR. *Journal of Petroleum Science Engineering* 3(3): 309–320. https://doi.org/10.1016/j.petlm.2017.03.003.

Verma, M.K. 2015. Fundamentals of carbon dioxide enhanced oil recovery (CO_2-EOR): A supporting document of the assessment methodology for hydrocarbon recovery using CO_2-EOR associated with carbon sequestration. *U.S. Geological Survey* 15: 1–19. https://doi.org/10.3133/ofr20151071.

Vledder, P., Fonseca, J.C., Wells, T., Gonzalez, I. and Ligthelm, D. 2010. Low Salinity Water Flooding: Proof of Wettability Alteration on a Field Wide Scale. *Presented at the SPE Improved Oil Recovery Symposium*, 24–28. https://doi.org/10.2118/129564-MS.

Webb, K.J., Black, C.J.J. and Al-Ajeel, H. 2004. Low Salinity Oil Recovery Log-Inject-Log. *Paper presented at the SPE/DOE Symposium on Improved Oil Recovery*, 17–21. https://doi.org/10.2118/89379-MS.

Wu, D., Brantson, E.T. and Ju, B. 2021. Numerical simulation of water alternating gas flooding (WAG) using CO_2 for high-salt argillaceous dolomite reservoir considering the impact of stress sensitivity and threshold pressure gradient. *Acta Geophysica* 69(4): 1349–1365. https://doi.org/10.1007/s11600-021-00601-w.

Yang, D., Tontiwachukwuthikul, P. and Gu, Y. 2005. Interfacial tensions of the crude oil + reservoir brine + CO_2 systems at pressures up to 31 MPa and temperatures of 27°C and 58°C. *Journal of Chemical Engineering Data* 50(4): 1242–1249. https://doi.org/10.1021/je0500227.

Yong, T., Zhengywan, S., Jibo, H. and Fulin, Y. 2016. Numerical simulation and optimization of enhanced oil recovery by the *in situ* generated CO_2 huff-n-puff process with compound surfactant. *Journal of Chemistry* 206: 13. https://doi.org/10.1155/2016/6731848.

Zekri, A., Al-Attar, H., Al-Farisi, O., Almehaideb, R. and Lwisa, E.G. 2015. Experimental investigation of the effect of injection water salinity on the displacement efficiency of miscible carbon dioxide WAG flooding in a selected carbonate reservoir. *Journal of Petroleum Exploration and Production Technology* 5: 363–373. https://doi.org/10.1007/s13202-015-0155-0.

Zolfaghari, H., Zebarjadi, A., Shahrokhi, O. and Ghazanfari, M.H. 2013. An experimental study of CO_2-low salinity water alternating gas injection in sandstone heavy oil reservoirs. *Iranian Journal of Oil and Gas Science and Technology* 2(3): 37–47. https://doi.org/10.22050/ijogst.2013.3643.

CHAPTER 8

Improving Seismic Salt Mapping through Transfer Learning Using A Pre-trained Deep Convolutional Neural Network
A Case Study on Groningen Field

Daniel Asante Otchere,[1,2,]* *Abdul Halim Latiff,*[1] *Nikita Kuvakin,*[3] *Ruslan Miftakhov,*[3] *Igor Efremov*[3] and *Andrey Bazanov*[3]

1. Introduction

The interpretation of seismic data is a critical aspect of geological exploration. It enables geologists and engineers to identify and delineate subsurface structures such as faults, reservoirs, and geological formations. However, traditional approaches to seismic interpretation, such as manual picking and horizon tracking, are laborious and time-consuming, particularly in areas with complex geology and numerous faults. Furthermore, these methods are prone to noise and other stratigraphic challenges, making them less reliable and accurate (Otchere et al., 2022c). Getting a reliable velocity model using

---

[1] Centre of Research for Subsurface Seismic Imaging, Universiti Teknologi PETRONAS, 32610, Seri Iskandar, Perak Daril Ridzuan, Malaysia.
[2] Institute for Computational and Data Sciences, Pennsylvania State University, University Park, PA, USA.
[3] GridPoint Dynamics, 77 Hopton Road, London, SW162EL, United Kingdom.
* Corresponding author: ascotjnr@yahoo.com

seismic salt imaging requires consistent interpretation analysis. However, the following restrictions may apply to traditional methods:

1. Clearly delineated salt boundaries at lithological transition zones.
2. Mapping complex salt build-ups require a great deal of time.
3. Instead of using automatic picking, a significant amount of manual horizon interpretation is required.

Salts are essential subsurface structures that significantly impact the geology of a region, and salt tectonics play a critical role in the storage of CO_2 gas, which is an essential part of the global energy transition journey (Duffy et al., 2022). The study of salts and their geological properties is of immense importance for the energy industry and our understanding of the Earth's subsurface processes. The application of deep learning has revolutionised the oil and gas industry in the last decade, providing a new and efficient approach to seismic interpretation (Otchere et al., 2022c), reservoir engineering (Otchere et al., 2022a), and subsurface characterisation (Otchere et al., 2022b). This technology has significantly contributed to mapping and imaging subsurface structures, particularly salt structures, which are notoriously difficult to interpret due to their complex geometries and stratigraphic complexities. Compared to traditional seismic interpretation methods, deep learning-based algorithms offer several advantages, such as higher accuracy and faster processing times. Additionally, the ability of deep learning models to learn from large datasets and identify complex patterns makes them ideal for identifying subtle subsurface features that conventional techniques may miss. Consequently, applying deep learning-based algorithms in the oil and gas industry has grown tremendously in recent years. It is expected to continue to play a significant role in future subsurface characterisation, interpretation, and modelling.

Seismic edge-detection algorithms have been widely used in the oil and gas industry to detect and interpret stratigraphic features in seismic data volumes. These algorithms have become increasingly popular due to their ability to extract subtle features in seismic data that may be indicative of subsurface structures (Chopra and Marfurt, 2007). They are particularly useful in areas where the subsurface is complex and difficult to interpret, such as in areas with numerous faults or where stratigraphic layering is intricate. Furthermore, edge-detection algorithms have been successfully integrated with machine learning techniques to improve the accuracy of interpretation results and reduce the time required for interpretation (Otchere et al., 2022c). These advancements have opened up new possibilities for the efficient and effective mapping and characterisation of subsurface structures, with significant implications for the oil and gas industry. However, deploying the method to new fields gives inaccurate mapping interpretation. One way of boosting deep learning model performance in a new field is through transfer learning.

In recent years, there has been a growing interest in using transfer learning to improve the accuracy of salt segmentation in seismic images. Transfer learning can significantly enhance our ability to identify and map subsurface salt structures, which can significantly impact the development of new energy resources and climate change mitigation. Using the specialised information learned in one context to solve the problems in another is known as transfer learning, which is a promising approach to address disparities in model prediction and ground truth (Li et al., 2020; Pan and Yang, 2010). The idea is to get insights into a problem using a plethora of available interpreted data when the initial model prediction of a comparable situation for which less information is not desirable. This calls for attention to the distinction between the model prediction and ground truth, reflected in the different distributions of their features and boundaries. In particular, the feature-based transfer learning strategy known as "domain adaptation" is a potential option (Ben-David et al., 2007; Pan et al., 2011). To do this, it first determines a feature space in which source domain data retain their intrinsic structure while minimising distribution disparity across domains. Transfer component analysis is a pioneering technique described by (Pan et al., 2011). Minimisation of maximum mean discrepancy (MMD) has emerged as a popular method for domain adaptation in machine learning. This method aims to learn transferable features between different domains while preserving the variation in the source domain. By representing the feature space as a reproducing kernel Hilbert space (RKHS), MMD minimisation enables the computation of distances between probability distributions, facilitating knowledge transfer from the source to the target domain. In this way, MMD-based domain adaptation has been successfully applied to various problems, including image classification, object detection, and natural language processing. An implementation with multiple kernels has been developed using this method (Lixin Duan et al., 2012).

Deep learning models have revolutionised the field of machine learning by enabling computers to automatically learn complex representations of data. The ability to learn features from data has made deep learning particularly useful in domains where large amounts of data are available, such as computer vision and natural language processing. Recently, there has been an increasing interest in combining deep learning with transfer learning, which aims to leverage knowledge from related tasks to improve performance on new tasks. To this end, Ghifary et al. (2014) introduced a deep adaption neural network (DaNN) that includes a multi-modal dependency (M-MD) term in the loss function to minimise MMD while maintaining variation in the source domain. This approach has shown promising results in various applications, including image classification and speech recognition.

Domain adaptation has become an increasingly popular area of research in recent years as more and more problems arise with data distribution differences between the source and target domains. Among the various techniques developed to address these issues, domain-adversarial neural networks (DANN) have shown great potential (Ganin et al., 2016). The key idea behind DANN is to build a network that can predict the source domain problem while avoiding the discrimination between the source and target domains. This is achieved by using an adversarial loss function that helps minimise the difference between the feature representations of the source and target domains. The use of DANNs has been demonstrated in several applications, including image recognition, natural language processing, and speech recognition.

Shi et al. (2019) defined the challenge of interpreting and extracting salt boundaries as a 3D image segmentation problem and assessed a deep CNN technique using an encoder-decoder architecture. The authors designed a data generator to train the model that extracts randomly positioned sub-volumes from a large-scale 3D training data set, followed by data reinforcement. The binary classification of sub-volumes into salt and non-salt was used as ground truth labels by thresholding the velocity model. These were then fed into the network, resulting in numerous iterations of network training. This approach enabled the network to learn salt features effectively, leading to more accurate salt boundary detection in seismic volumes. Their model, when validated, automatically extracted minor salt properties from 3D seismic sections. Liu et al. (2019) also used a Squeeze-Extraction Feature Pyramid Network (Se-FPN) for salt deposits image segmentation. They used SeNet as a backbone to implicitly learn to suppress unimportant regions while highlighting key features. The authors suggested an enhanced FPN to combine multiscale information, where hypercolumns were added to the network to combine information from multiple scales. The proposed Se-FPN, when deployed in the TGS Salt Identification Challenge, got high-quality segmentation with a Mean Intersection over Union (MIU) of 0.86. To overcome the discrepancies between model fault prediction and ground truth (Yan et al., 2021), used transfer learning to improve the prediction performance of a CNN pre-trained with synthetic labels. The authors retrained the model using actual field seismic interpretations. They used random sample consensus (RANSAC) to automatically obtain and categorise actual seismic samples. This approach resulted in an improvement in fault detection accuracy. The existing literature provides a theoretical foundation for our objective, which is to leverage transfer learning to enhance the detection of salt bodies and evaluate the effectiveness of the pre-trained CNN. These prior investigations demonstrate that transfer learning can be used to improve the accuracy of CNNs for image recognition tasks, even when the target dataset differs from the source dataset. The domain shifts (actual-predicted discrepancy) between the predicted and ground truth can be reduced to an acceptable level in a predetermined feature space using domain adaptation through transfer learning.

This study will demonstrate the application of transfer learning using actual field interpretation to a CNN pre-trained with synthetic labels to generate salt probability models that can be used as a valuable property in the seismic imaging and velocity modelling phases. Transfer learning and deep learning techniques' object and edge detection have demonstrated promising success in various domains, making them an appealing approach for seismic salt mapping. The use of these techniques can potentially enhance the detection of subsurface salt bodies and improve the accuracy of salt mapping, resulting in more efficient and effective exploration and production of hydrocarbon reservoirs. Hence, the main contributions of this research are:

1. *Improved salt segmentation accuracy*: Transfer learning can significantly enhance our ability to identify and map subsurface salt structures in seismic images. Using semantic segmentation to improve salt probability volume by retraining the model on a labelled dataset, it can learn to recognise salt structures with greater accuracy and efficiency, which is critical for the energy industry and our understanding of the Earth's subsurface processes.

2. *Cost and time savings*: Salt segmentation in seismic images is time-consuming and labour-intensive, requiring significant expertise and resources. By using transfer learning to automate this process, we can save time and reduce costs while improving the accuracy of the segmentation.

3. *Enhanced data analysis*: Accurate segmentation of salt structures in seismic images is essential for various applications, including hydrocarbon exploration, CO_2 storage, and geohazard assessment. By improving the accuracy of salt segmentation, we can enhance our understanding of the subsurface geology and make more informed decisions about energy exploration and production.

4. *Potential for future research*: Using transfer learning for salt segmentation in seismic images is a relatively new area of research with significant potential for future exploration and development. By demonstrating the effectiveness of this approach, we can inspire new research in the field and help to advance our understanding of subsurface processes and the energy transition journey.

These contributions will help reduce the time and resources seismic interpreters spend interpreting seismic features. The findings in this study are of great importance for the energy industry and the broader scientific community, as they provide new insights into the use of transfer learning for improving the accuracy of salt segmentation in seismic images. The results of this study will also highlight the potential of this approach for developing new energy resources and mitigating climate change by improving our ability to identify and map subsurface salt structures accurately.

2. Method

2.1 Collection and Description of Data

For this work, the Groningen seismic field data was selected as the case study. The system unit was a Windows 10 Operating System (OS) with the next generation AMD Radeon Pro™ and the highest-performing NVIDIA Quadro® professional graphics capable of 2-petaFLOPS tensor performance. A single NVIDIA RTX 8000 GPU and quadruple Intel (R) Xeon (R) W-2223 i5 CPU running at 3.60 GHz, with 32.0 GB of DDR4-2666 MHz DRAM, were used to run the study. This research used a pre-trained DCNN model trained using synthetic and real data. The proposed approach consisted of two main parts, generating a salt probability model and improving the model output through transfer learning, as illustrated in Fig. 8.1. The optimised DCNN model for salt body probability prediction can be reused to map new salt bodies, thus enhancing interpretation efficiency. The model estimation section of the workflow represents an end-to-end model that automatically can map salt bodies after selecting several parameters.

2.2 Deep Convolutional Neural Network in Salt Mapping and Post-processing

For a long time, seismic data has become increasingly important in oil and gas exploration due to the valuable information it provides about subsurface structures. Seismic data is usually obtained by sending a wave signal into the ground and measuring the reflected signals that bounce back from different subsurface layers. This data is processed and interpreted to create seismic images of the subsurface structures. With the advancement in exploration technology, 3D seismic data volumes have become more common, providing a more detailed and accurate representation of the subsurface geology. However, the large volume of data generated by this technology presents a significant challenge for interpretation and analysis, necessitating the development of more efficient and effective techniques. To reduce the computation cost associated

Improving Seismic Salt Mapping through Transfer Learning 165

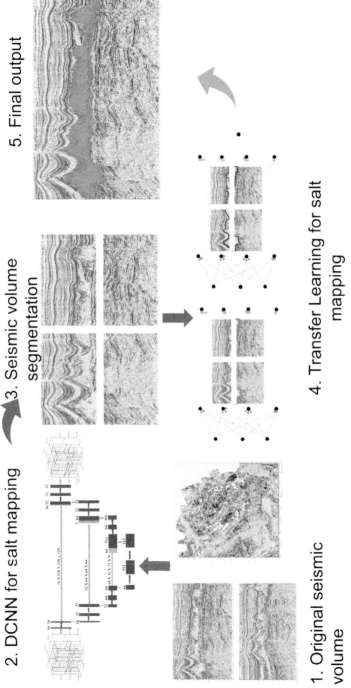

Fig. 8.1. Illustration of workflow using a pre-trained transfer learned DCNN seismic salt mapping model.

with training a neural network directly on 3D seismic data, the Geoplat AI software was employed in this study to import the entire 3D seismic volume and the interpreted salts of the Groningen field. The software's pre-trained DCNN, which is based on a computer vision-based model, was used. The model was trained on an extensive and diverse 3D synthetic seismic dataset image library created by fragmenting labelled images. These synthetic data were carefully designed to capture complex geological factors and imaging quality. These images were segmented using the DCNN model, which was initially trained to assign a label of 1 to salt-labelled images and a label of 0 to non-salt-labelled images, as illustrated in Fig. 8.2. The accurate segmentation of a synthetic volume fragment is shown on the left, while the predicted output created by the pre-trained neural network is shown on the right.

The pre-trained DCNN model's outstanding performance prompted its application to the Groningen Field's actual dataset for automatic salt probability volume creation. The selected fragment size for the prediction was based on visually identifying salt tops and bottoms in the seismic volume. The resultant inline, crossline, and depth sections of the Groningen Field are displayed in Fig. 8.3. The ResU-net architecture framework, which has been pre-trained, is used to segment the 3D seismic images obtained from the seismic volume. Figure 8.4 illustrates the application framework of the DCNN for the creation of the salt probability volume. An additional signal gain was applied to handle low-quality seismic data, which amplifies and restores the regions with amplitude expression coefficient loss.

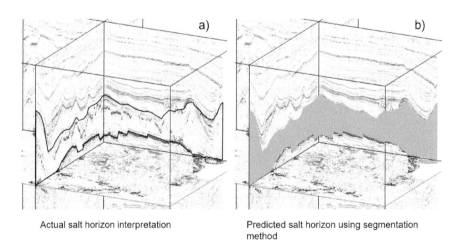

Actual salt horizon interpretation   Predicted salt horizon using segmentation method

Fig. 8.2. (a) Accurate segmentation of a synthetic volume fragment, and (b) projected result produced using a pre-trained neural network.

Improving Seismic Salt Mapping through Transfer Learning   167

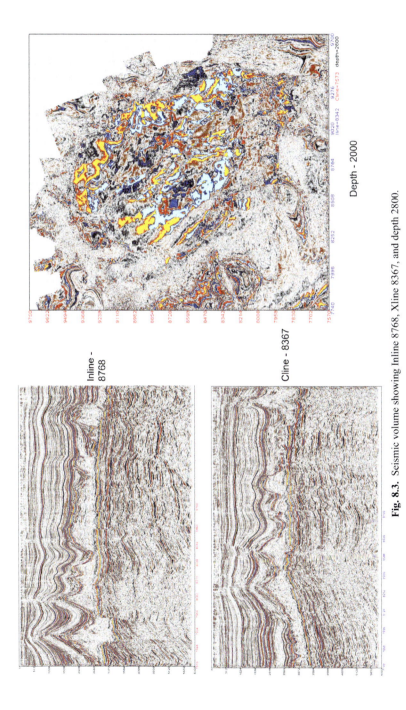

Fig. 8.3. Seismic volume showing Inline 8768, Xline 8367, and depth 2800.

168 Data Science and Machine Learning Applications in Subsurface Engineering

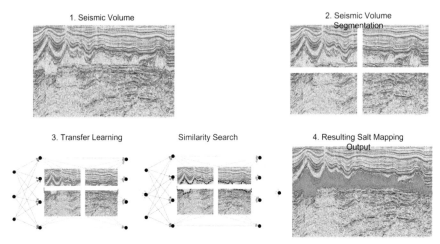

Fig. 8.4. Illustration of the application framework of the DCNN in creating the salt probability volume.

2.2.1 Simplified Architecture of Residual U-net

The ResU-net architecture used in this study serves as the backbone of the DCNN architecture employed. The ResU-net capitalises on the power of deep residual learning to overcome encoder challenges by transferring information between layers, resulting in a deep neural network that effectively deals with degradation issues. Figure 8.5 depicts a simplified schematic illustration of the ResU-net, which utilises complete pre-activation deep residual units to train the model. Using bypass connections inside networks aids information propagation without degradation, enhancing the convolutional neural network's architecture by reducing parameters while maintaining or improving the transfer learning function's efficiency.

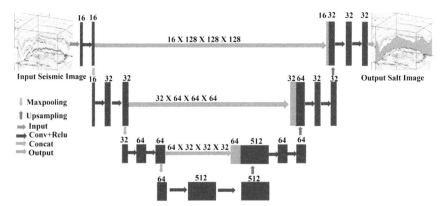

Fig. 8.5. A simplified end-to-end U-network architecture for 3D salt mapping.

2.3 Transfer Learning Application

Semantic segmentation can help improve probability findings tailored to the interpretation interval's peculiarities. The failure of the default probability volume to deliver quantitative results underpins the necessity for the transfer learning technique. Concerning post-processing, there is the possibility of fine-tuning the DCNN to handle the problem of thorough tracking of faults peculiar to a specific section of the seismic area. The DCNN allows it to address the difficulty of automating the interpretation of highly complicated faults. Hence, expert transfer learning based on manually interpreted salt horizons was incorporated to improve the accuracy of the prediction in places where there is minimal definitive evidence of salt and also to include the existing interpreted salts in the Groningen Field.

In applying the transfer learning process, one set of salt body interpretations on Inlines 9090, 7820, and 9108 was used as the training segment. In contrast, a different set on Inline 8271 was used as the validation section (Fig. 8.6). A different set of interpreted salt on crosslines 7960, 9034, and 8188 was used as part of the transfer learning process, with the latter used for validation (Fig. 8.7). The training and validation segments were chosen irrespective of each other and ease of interpreting salt horizons or otherwise. After selecting the interpreted salt sections for the transfer learning process, several iterations and output weights were specified with the results compared. The iterations and weights are systematically chosen based on the erroneous probabilities in the probability volume created using new weights and the dice coefficient value on training and validation segments. Where underfitting is observed, the number of iterations is increased to improve the model performance whiles also critically avoiding overfitting.

2.4 Criteria for Model Evaluation

In assessing the performance of the DCNN predicted salt bodies, the models' errors were assessed using the Dice Similarity Coefficient (DSC), commonly known as the Sørensen–Dice index, is a numerical tool based on spatial overlap used for determining how similar two sets of data are. This index is likely the most widely used tool for validating artificial intelligence-based (AI) image segmentation predictions. The validation challenge becomes an assessment of segmentation reproducibility and spatial overlap index. It was sometimes referred to as the percentage of explicit agreement (Zou et al., 2004). DSC values range from 0 to 1, with 0 indicating no spatial overlap and 1 indicating

170 *Data Science and Machine Learning Applications in Subsurface Engineering*

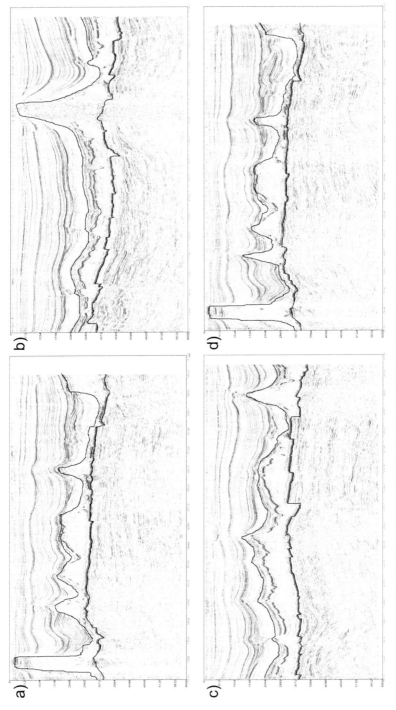

Fig. 8.6. Expert interpreted salt bodies showing: (a) Inline 9090, (b) Inline 7820, (c) Inline 8271 and (d) Inline 9108.

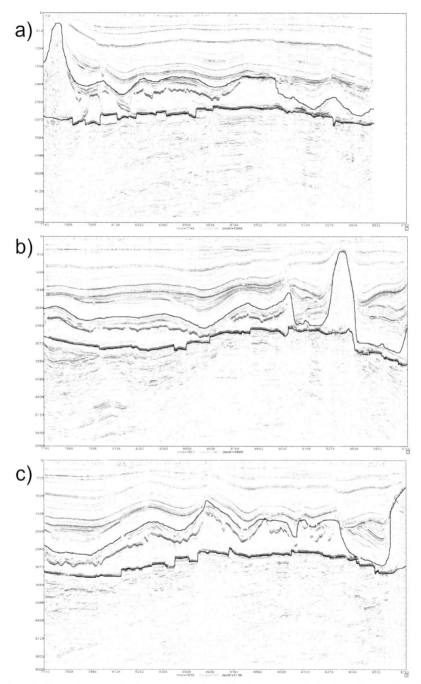

Fig. 8.7. Expert interpreted salt bodies showing: (a) Crossline 9034, (b) Crossline 7960, and (c) Crossline 8188.

172 *Data Science and Machine Learning Applications in Subsurface Engineering*

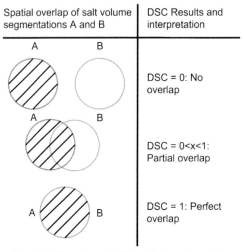

Fig. 8.8. Illustration of DSC spatial overlap metric.

complete overlap between two sets of binary image segmentation results, A and B target regions (Fig. 8.8). The DSC is defined as (Yao et al., 2020);

$$DSC(A, B) = 2(A \cap B)/(|A| + |B|) \quad (8.1)$$

where $\cap$ represents the intersection of the two target regions |A| and |B| is the cardinality of the set. When there is a much higher number of background voxels than the number of target voxels, the DSC can be conceptualised as a special case of the kappa statistic, typically applied in situations involving reliability analyses. This concept has been shown previously by Zou et al. (2004).

3. Results and Discussion

3.1 Calculated Salt Body Volume

In Fig. 8.9, the low level of accuracy in predicting salt probability bodies is noted. However, the possibility of using DCNN to map salt bodies on the base of seismic data has been illustrated. This low accuracy predicted salt probability volume could give an indication of the location and boundaries of the salt bodies. While the calculated salt body volume provides a broad indication of the presence of salt in the Groningen Field, it does not provide precise information regarding the exact position of the top, base, and wall sides of salt bodies. This is because the calculated volume is based on the probabilistic output of the DCNN model. As such, it may contain sections of "false probability" where the analysis of the reflectivity pattern does not suggest the presence of a salt body.

Improving Seismic Salt Mapping through Transfer Learning 173

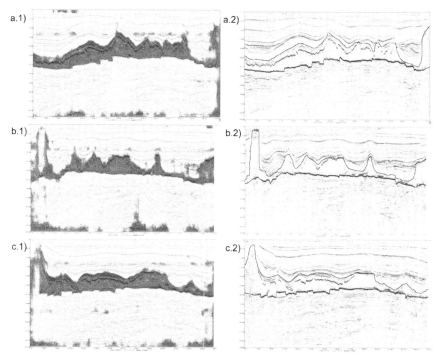

Fig. 8.9. Comparison of predicted salt probability volume (left) and expert interpretation (right) of (a) Crossline 8188, (b) Inline 9090, and (c) Crossline 9034.

Nonetheless, these areas are still shown on the attribute as having a high probability of salt occurrence, indicating the potential presence of salt. Further analysis and interpretation by geoscientists are required to accurately determine the position and extent of salt bodies within the field. This event is because "false probability" sections occur when there is no obvious evidence of a salt body. The occurrence of probabilities, which are bodies traced in a distinct direction, is projected to have a salt probability. These probabilities are a fragmented portrayal of some extensive salt bodies and the requirement for vertical extent rectification. These false probabilities occur because synthetic and field labels used to pre-trained DCNN do not capture all the intrinsic characteristics in every field. Some of these false probabilities are salt probability zone projections tracked in the perpendicular direction. One approach for enhancing the probability attribute in such cases is by adding manually interpreted salt labels to retrain the pre-trained DCNN. Due to the unsatisfactory salt body prediction by the pre-trained model, transfer learning is recommended to help improve the model performance by tuning and recalculating the model weights to the type of salt bodies in this study. This approach will decrease the number of "false probability" results where the analysis of the reflectivity pattern does not suggest salt being present.

174 *Data Science and Machine Learning Applications in Subsurface Engineering*

3.2 Semantic Segmentation – Transfer Learning Application

A high similarity between the transfer learned salt body detection and the interpreted salt bodies in the seismic sections could be noticed in the final salt probability volume, as shown in Figs. 8.11 and 8.10. The neural network was

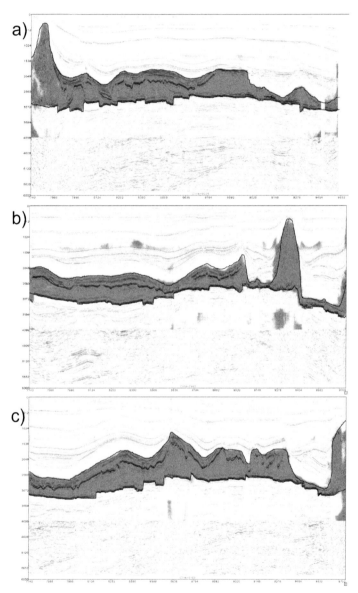

Fig. 8.10. Transfer learned salt probability volume and expert interpretation (solid black line) showing: (a) Crossline 9034, (b) Crossline 7960, and (c) Crossline 8188.

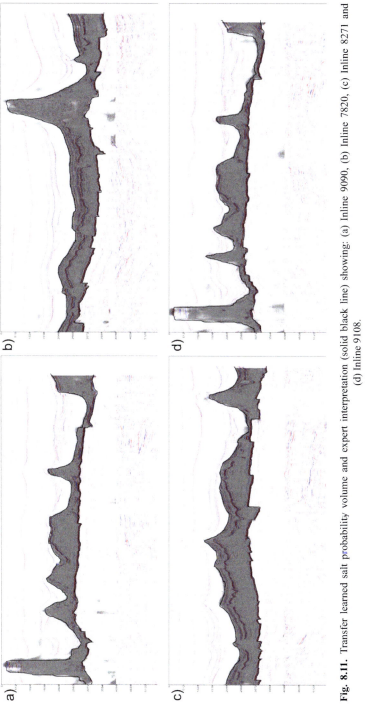

Fig. 8.11. Transfer learned salt probability volume and expert interpretation (solid black line) showing: (a) Inline 9090, (b) Inline 7820, (c) Inline 8271 and (d) Inline 9108.

able to successfully retrain itself on the new labelled data set by readjusting its weights in the deep learning process. As a result, the accuracy in the prediction of salt probability bodies has increased compared to the initial results obtained before the retraining of the CNN. The results showed a statistically significant improvement when iterations were increased in steps of 100 from 100 to 10,000. The best combination of input data and iterations was determined by comparing the pair-wise DSC metrics of all the iterations and increasing the number of input labelled data. DSC results sequentially increased with successive increments in labelled training data and iterations until they reached their maximum based on case-to-case analysis. There was no evidence of a learning curve phenomenon, and segmentations seemed to have little effect after iterations 10,000 and input labelled data set size above 5.

This research demonstrated the functionality of DSC as a straightforward criterion for validating the consistency and spatial overlap accuracy of manual and automated segmentations. The corresponding DSC values for iteration 10,000 using seven input labelled data for training and validation (four inlines and three crosslines) are shown in Fig. 8.12. The probabilistic fractional segmentation showed a wide range of spatial overlap (0.6–0.95) with the corresponding estimated ground truth (expert interpreted sections). This result gives an indication that further improvements can be achieved. In the inline sections, the reproducibility was significantly higher based on a higher number of labelled training inputs. This improvement motivated the increment of the labelled data size in subsequent crossline investigations to develop a suitable matching approach to register these salt body boundaries

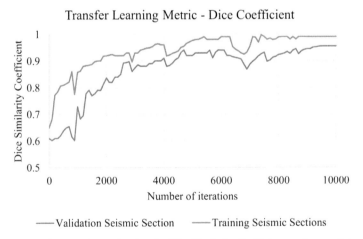

Fig. 8.12. Learning transfer metrics showing Dice Similarity Coefficient results at each iteration.

and enhance the visual representation of salt body mapping. However, increasing the training labelled data for the crossline section did not yield any significant improvements.

3.3 Sensitivity Analysis of Model and Expert Interpretations

Image segmentation is a critical task in many fields, including geophysics, where it can be used to identify and extract features of interest from seismic data. The accuracy and consistency of segmentation results are essential for ensuring the reliability of subsequent analysis and interpretation. One approach to improving segmentation performance is to use transfer learning, which involves fine-tuning a pre-trained CNN on new data. The transfer learned CNN based on the statistical validation analysis based on DSC was expanded to different seismic sections from the same field compared to the ground truth of expert interpretations. The results in Fig. 8.13 indicate no significant difference in DSC between the predicted salt probability volume and manual interpretation. This result suggests that the repeated segmentations through transfer learning did not introduce significant bias. This observation is an important finding because it indicates that transfer learning can be used to effectively extend a model's performance to new data without compromising its accuracy or consistency.

To highlight the critical points for reinterpretations, we recommend that future studies focus on the following areas:

1. Transfer learning can be a practical approach for extending the performance of a segmentation model to new data.
2. The dice similarity index can be a valuable metric for evaluating segmentation performance. However, it is recommended to include other metrics, like Intersection over Union (IoU), to confirm the final results.
3. The use of transfer learning did not introduce significant bias into the segmentation results.
4. Further validation studies can help confirm segmentation results' reliability and consistency using transfer learning.

In summary, our study demonstrates the potential of transfer learning for improving the accuracy and consistency of image segmentation in geophysics. Using the DSC as our evaluation metric, we validated the performance of our transfer learned CNN on new data and confirmed the absence of significant bias. These findings provide valuable insights for future research and the development of automated geophysical segmentation tools.

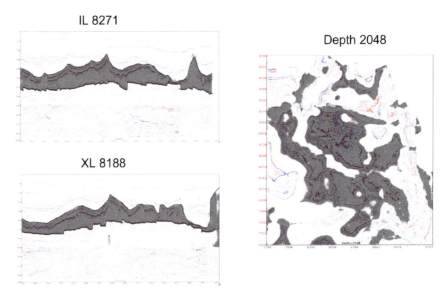

Fig. 8.13. Transfer learned salt probability volume and expert interpretation (solid black line) showing the deployed sections of Inline 8271, crossline 8188 and depth 2048.

4. Conclusions

The seismic interpretation field has faced significant challenges due to the limited availability of labelled seismic data. These challenges have created a bottleneck in developing deep learning algorithms to aid seismic interpretation. Accurate and high-quality seismic data are critical to successfully applying machine learning algorithms to interpret subsurface geological structures. Without sufficient labelled data, deep learning models are unable to learn the patterns and correlations necessary to accurately predict subsurface features, such as salt bodies. In order to solve this issue, a CNN that had previously been trained on synthetic labels was considered a potential source of relevant data for developing an adequate salt segmentation model. The application of transfer learning was required to reduce the visual discrepancy across the seismic volume because of the inevitable disparity between the model and the actual data. The application of this process in the Groningen field has demonstrated the effectiveness of integrating DNNs and transfer learning for salt body detection. In its conceptualisation, DSC is a particular instance of the kappa statistic, a well-known and pragmatic reliability and agreement index. It offers the metric to analyse and measure the model's performance and the adjustments necessary for the retraining process. The transfer learning process might involve several repetitions. Depending on the data quality and the geological complexity, it may be necessary to complete several iterations to achieve a representative model for the deployed seismic field. The transfer learning technique in this study achieved a Dice similarity index of 0.99 and 0.92 on the training and validation sections, respectively. This result shows that the transfer learned model can automatically capture subtle salt bodies from 3D seismic with minimal manual input.

The proposed method advocates for continued research and deep learning techniques implementation in seismic interpretation. This activity involves manual seismic interpretation and often requires much expert knowledge. It would be beneficial to have labelled data of interpreted seismic from different fields. Despite a large portion of training data coming from synthetic data, the successful knowledge transfer still requires some field interpretation samples from actual seismic volumes. Once labelled data is made available, the time involved in fully interpreting salt boundaries or other attributes on seismic volumes will be drastically reduced because of the generalisation of the DCNN model and can accurately map and interpret different seismic volumes.

Data and Software Availability

The data used in this study is available at https://geo.public.data.uu.nl/vault-nam-geological-model/Publication%5B1605778324%5D/original/. To access the software used in this article, contact the developers at https://geoplat.ai/#demo.

Acknowledgement

The authors wish to extend their genuine gratitude to the University Teknologi Petronas and the Centre for Subsurface Seismic Imaging for their support and valuable contributions to this research and to Geoplat AI for providing the software for this work.

References

Ben-David, S., Blitzer, J., Crammer, K. and Pereira, F. 2007. Analysis of representations for domain adaptation. pp. 137–144. *In*: *Advances in Neural Information Processing Systems 19*. The MIT Press. https://doi.org/10.7551/mitpress/7503.003.0022.

Chopra, S. and Marfurt, K.J. 2007. Volumetric curvature attributes for fault/fracture characterisation. *First Break* 25: 35–46. https://doi.org/10.3997/1365-2397.2007019.

Duffy, O.B., Hudec, M., Peel, F., Apps, G., Bump, A., Moscardelli, L., Dooley, T., Bhattacharya, S., Wisian, K. and Shuster, M. 2022. *The Role of Salt Tectonics in the Energy Transition: An Overview and Future Challenges*. Earth Arxiv Preprint. https://doi.org/10.31223/X5363J.

Ganin, Y., Ustinova, E., Ajakan, H., Germain, P., Larochelle, H., Laviolette, F., Marchand, M. and Lempitsky, V. 2016. Domain-adversarial training of neural networks. *Advances in Computer Vision and Pattern Recognition* 189–209. https://doi.org/10.1007/978-3-319-58347-1_10.

Ghifary, M., Kleijn, W.B. and Zhang, M. 2014. Domain adaptive neural networks for object recognition. pp. 898–904. *In*: *Pacific Rim International Conference on Artificial Intelligence*. Springer, Cham. https://doi.org/10.1007/978-3-319-13560-1_76.

Li, W., Gu, S., Zhang, X. and Chen, T. 2020. Transfer learning for process fault diagnosis: Knowledge transfer from simulation to physical processes. *Comput. Chem. Eng.* 139: 106904. https://doi.org/10.1016/j.compchemeng.2020.106904.

Liu, B., Jing, H., Li, J., Li, Y., Qu, G. and Gu, R. 2019. Image segmentation of salt deposits using deep convolutional neural network. pp. 3304–3309. *In*: *2019 IEEE International Conference on Systems, Man and Cybernetics (SMC)*. IEEE. https://doi.org/10.1109/SMC.2019.8913858.

Lixin Duan, Tsang, I.W. and Dong Xu. 2012. Domain transfer multiple kernel learning. *IEEE Trans. Pattern Anal. Mach. Intell.* 34: 465–479. https://doi.org/10.1109/TPAMI.2011.114.

Otchere, D.A., Abdalla Ayoub Mohammed, M., Ganat, T.O.A., Gholami, R. and Aljunid Merican, Z.M. 2022a. A novel empirical and deep ensemble super learning approach in predicting reservoir wettability via well logs. *Applied Sciences* 12: 2942. https://doi.org/10.3390/app12062942.

Otchere, D.A., Ganat, T.O.A., Nta, V., Brantson, E.T. and Sharma, T. 2022b. Data analytics and Bayesian Optimised Extreme Gradient Boosting approach to estimate cut-offs from wireline logs for net reservoir and pay classification. *Appl. Soft Comput.* 120: 108680. https://doi.org/10.1016/j.asoc.2022.108680.

Otchere, D.A., Tackie-Otoo, B.N., Mohammad, M.A.A., Ganat, T.O.A., Kuvakin, N., Miftakhov, R., Efremov, I. and Bazanov, A. 2022c. Improving seismic fault mapping through data conditioning using a pre-trained deep convolutional neural network: A case study on Groningen field. *J. Pet. Sci. Eng.* 213: 110411. https://doi.org/10.1016/J.PETROL.2022.110411.

Pan, S.J. and Yang, Q. 2010. A survey on transfer learning. *IEEE Trans. Knowl. Data Eng.* 22: 1345–1359. https://doi.org/10.1109/TKDE.2009.191.

Pan, S.J., Tsang, I.W., Kwok, J.T. and Yang, Q. 2011. Domain adaptation via transfer component analysis. *IEEE Trans. Neural Netw.* 22: 199–210. https://doi.org/10.1109/TNN.2010.2091281.

Shi, Y., Wu, X. and Fomel, S. 2019. SaltSeg: Automatic 3D salt segmentation using a deep convolutional neural network. *Interpretation* 7: SE113–SE122. https://doi.org/10.1190/INT-2018-0235.1.

Yan, Z., Zhang, Z. and Liu, S. 2021. Improving performance of seismic fault detection by fine-tuning the convolutional neural network pre-trained with synthetic samples. *Energies (Basel)* 14: 3650. https://doi.org/10.3390/en14123650.

Yao, A.D., Cheng, D.L., Pan, I. and Kitamura, F. 2020. Deep learning in neuroradiology: a systematic review of current algorithms and approaches for the new wave of imaging technology. *Radiol. Artif. Intell.* 2: e190026. https://doi.org/10.1148/ryai.2020190026.

Zou, K.H., Warfield, S.K., Bharatha, A., Tempany, C.M.C., Kaus, M.R., Haker, S.J., Wells, W.M., Jolesz, F.A. and Kikinis, R. 2004. Statistical validation of image segmentation quality based on a spatial overlap index1. *Acad. Radiol.* 11: 178–189. https://doi.org/10.1016/S1076-6332(03)00671-8.

CHAPTER 9

Super-Vertical-Resolution Reconstruction of Seismic Volume Using A Pre-trained Deep Convolutional Neural Network
A Case Study on Opunake Field

Daniel Asante Otchere,[1,2,][*] *Abdul Halim Latiff*,[1] *Nikita Kuvakin*,[3] *Ruslan Miftakhov*,[3] *Igor Efremov*[3] and *Andrey Bazanov*[3]

1. Introduction

Seismic surveys are essential for subsurface structure exploration and analysis in the oil and gas and civil engineering sectors (Farfour et al., 2021; Talwani and Kessinger, 2003). Reflection seismic surveys are acquired to map potential subsurface features and structural and stratigraphic hydrocarbon traps (Haldar, 2018). The subsurface rock's acoustic impedance changes produce reflections due to the Earth's reaction to synthetically generated acoustic waves near the surface (Aminzadeh and Dasgupta, 2013; Selley and Sonnenberg, 2015). Although the seismic energy travels in the form of elastic waves, most imaging procedures presume that the recorded waves are exclusively composed of compressional waves (Gray, 2014). Seismic image resolution is the ability to accurately and delicately depict subsurface structures from the Earth's

---

[1] Centre of Research for Subsurface Seismic Imaging, Universiti Teknologi PETRONAS, 32610, Seri Iskandar, Perak Daril Ridzuan, Malaysia.
[2] Institute for Computational and Data Sciences, Pennsylvania State University, University Park, PA, USA.
[3] GridPoint Dynamics, 77 Hopton Road, London, SW162EL, United Kingdom.
* Corresponding author: ascotjnr@yahoo.com

reflected energy, as it determines the level of detail that can be observed in seismic images (Roy Chowdhury, 2011). However, several factors, including subsurface structure complexity, noise and interference in the data, low sensor resolution, seismic energy wavelength, and spatial sampling interval of the data acquisition system, frequently limit the resolution of seismic images (Ker et al., 2010). These constraints impede the efficiency of precisely interpreting the subsurface structures, resulting in costly exploration and production activity errors.

Artificial intelligence (AI) has been increasingly used in the oil and gas industry to improve its activities in subsurface characterisation, reservoir engineering, and production (Otchere et al., 2022a). Recent advances in AI, notably computer vision, have shown promise in enhancing seismic image quality and limiting the impact of noise and interference. Through learning patterns and features in the seismic images and using this knowledge to fill in missing or impaired sections of new images, AI approaches can be utilised to improve the spatial resolution of seismic images. These approaches have the potential to overcome conventional methods' limitations and offer more accurate and detailed images of subsurface structures.

Several studies have applied AI techniques to improve the resolution of seismic images to improve subsurface interpretation activities. For example, Li et al. (2022) used a deep convolutional neural network (DCNN) to enhance low-resolution and noisy seismic images, demonstrating improved image quality and resolution for fault detection. (Otchere et al., 2022b) employed a DCNN to improve the seismic resolution of the Groningen field, resulting in a more accurate interpretation of subsurface faults. Picetti et al. (2019) utilised a generative adversarial network (GAN) to process low-quality seismic migrated images and improve the resolution of features. These studies demonstrate the potential of AI in overcoming the limitations of traditional methods and improving the resolution of seismic images.

In summary, seismic image resolution plays a vital role in subsurface exploration and analysis. However, seismic imaging has inherent limitations regarding image resolution and quality, which might affect the accuracy and reliability of the resulting interpretation. AI techniques have the potential to improve seismic image resolution and quality using advanced machine learning algorithms, such as denoising and super-resolution techniques, and overcome the limitations of conventional techniques. Further research is thus required to optimise and validate the potential of AI completely and to explore its potential for automating the interpretation of seismic images.

This study will demonstrate the application of a pre-trained DCNN on the Opunake field for seismic image resolution improvement and to improve the signal-to-noise ratio (SNR) that can be used as a valuable property in the seismic imaging and velocity modelling phases. Two image resolution DCNN techniques will be compared to identify the most appropriate technique for vertical resolution improvements. The reported successes of using synthetic data to pre-train a DCNN due to the unavailability of data and deploying it on a real field make AI a viable candidate for vertical and lateral seismic image resolution.

2. Brief Overview

Seismic imaging is a well-known geophysical technique for visualising underlying structures and features in the Earth's subsurface. It is extensively used in the oil and gas sector to discover and map hydrocarbon reservoirs and in civil engineering for site characterisation and underground utility mapping. Despite its widespread use, Seismic imaging has intrinsic limitations in terms of image resolution and quality, which affects the accuracy and dependability of the ensuing interpretation. Several factors can limit the resolution of seismic images in the oil and gas industry (Deng et al., 2021; Sun et al., 2022; Yan and Chunqin, 2008):

1. *Seismic wavelength*: The resolution of a seismic image is directly related to the wavelength and finite size of the seismic source transmitted into the subsurface. Shorter wavelengths result in higher-resolution images, while longer wavelengths result in lower-resolution images.

2. *Seismic source frequency and depth*: The frequency of the seismic source (e.g., air gun, vibrator) also influences image resolution. Higher frequencies produce higher-resolution images but consume more energy and may not penetrate the subsurface as deeply. The resolution of a seismic image often decreases as depth increases due to the increased absorption of seismic energy as it passes through the subsurface.

3. *Geology*: The composition and structure of the Earth's subsurface can also affect the resolution of the seismic image. Harder, more consolidated materials produce clearer images than softer, more porous materials. Seismic energy can be scattered by complex geology, such as folds or faults, resulting in a lower-resolution image.

4. *Data processing*: The resolution of an image can be influenced by the quality of the data processing techniques employed. Image resolution can be improved using advanced processing techniques such as inversion and migration. Some techniques, such as filtering, can help to minimise noise and improve resolution, while others can reduce the resolution.

5. *Receiver spacing*: The spacing between the seismic receivers that collect data can also impact the image's resolution. Closer spacing results in higher-resolution images, but it also increases the cost of the survey.

Table 9.1. Summary of articles reviewed on seismic image resolution and image resolution using AI in oil and gas, civil engineering, and medical fields.

Author(s)	Title	Summary
\multicolumn{3}{c}{**Oil and Gas Field**}
Jo et al. (2022)	Machine learning-based vertical resolution enhancement considering the seismic attenuation	The authors used a Convolution U-net algorithm to enhance seismic image resolution. The researchers showed that using a spectral enhancement CNN U-net model trained with time-variant wavelets produced better outcomes for the seismically attenuated area, leading to a considerable increase in vertical image resolution.
Li et al. (2022)	Deep Learning for Simultaneous Seismic Image Super-Resolution and Denoising	The authors used CNN to improve the resolution of seismic images and denoise them. Compared to conventional methods, the CNN obtained better performance in enhancing detailed depiction of subsurface structural and stratigraphic structures.
Wang and Nealon (2019)	Applying machine learning to 3D seismic image denoising and enhancement	The authors utilised CNN to improve the resolution of marine seismic images for post-stack structural seismic image augmentation and noise reduction. They demonstrated that their method considerably improved seismic image resolution while preserving the varied scales of geologic structures, from high-resolution faults and diffractors to deep subsalt deposits, for better analysis.
Sun et al. (2022)	Random noise suppression and super-resolution reconstruction algorithm of seismic profile based on GAN	The authors recommended applying GANs to seismic profiles to improve low resolution and suppress random noise. Their efforts resulted in GANs obtaining an ideal SNR and significantly enhancing the resolution of the reconstructed cross-sectional seismic image, allowing for more vivid observations of geological structures such as fractures.
Lan et al. (2023)	Seismic Data Denoising Based on Wavelet Transform and the Residual Neural Network	This study employs a wavelet transform and a residual neural network (ResNet) to denoise and enhance seismic images. This approach demonstrated that the DWT-Resnet could isolate the effective signal from the noise using wavelet decomposition, which enhances the data quality going into the network and increases the resolution and accuracy, resulting in excellent noise suppression outcomes.
Orozco-del-Castillo et al. (2014)	A genetic algorithm for filter design to enhance features in seismic images	This study employs a genetic algorithm (GA) for kernel optimisation to improve the internal properties of salt bodies and the sub-salt stratigraphy. The GA was statistically compared to other algorithms and proved more appropriate for automatic interpretation.

Table 9.1 contd. ...

...Table 9.1 contd.

Author(s)	Title	Summary
\multicolumn{3}{c}{**Medical Field**}		
Cengiz et al. (2022)	Classification of breast cancer with deep learning from noisy images using wavelet transform	The authors suggested utilising the Wavelet Transform to denoise the input images and CNN to classify them. Based on PSNR measurements, the authors discovered that CNN produced good classification results.
Gondara (2016)	Medical Image Denoising Using Convolutional Denoising Autoencoders	Using a limited sample size, the author demonstrated that denoising autoencoders built using convolutional layers can effectively denoise medical images. The results show that heterogeneous images can be merged to enhance sample size for improved denoising performance.
Chen et al. (2018)	Brain MRI super-resolution using 3D deep densely connected neural networks	In this study, the authors used a 3D Densely Connected Super-Resolution Network (DCS) high-resolution elements of structural brain MR images. Using a dataset of 1,113 patients, they discovered that their neural network surpasses bicubic interpolation and other deep learning algorithms in reconstructing 4× resolution-reduced images. Including a CNN increased image resolution significantly and offered a more realistic depiction of anomalies in the scans.
Shi et al. (2019)	DeSpecNet: A CNN-based method for speckle reduction in retinal optical coherence tomography images	The use of a deep CNN to despeckle optical coherence tomography (OCT) images is proposed in this paper. The researchers discovered that their method resulted in significant improvements in visual quality and quantitative indices and high generalisation ability for various retinal OCT images. The CNN surpassed state-of-the-art approaches in suppressing speckles and showing subtle details while retaining edges.
Civil Engineering Field		
(Mukherjee et al., 2022)	High-resolution imaging of subsurface infrastructure using deep learning artificial intelligence on drone magnetometry	Using high-resolution aboveground magnetic data, the authors demonstrate the possibility of imaging pipelines and other subterranean infrastructure using AI-based methods. Their findings showed that their advanced algorithms and processes resulted in a 100-fold increase in efficiency compared to commonly available open-source deep learning AI workflows and software. Superior resolution and interpretability are also proven above the conventional geophysical inversion. The method could be extended to other geophysical data types at different scales and resolutions, such as seismic, electromagnetic, and gravity.

Table 9.1 contd. ...

...Table 9.1 contd.

Author(s)	Title	Summary
Jiang et al. (2019)	Edge-Enhanced GAN for Remote Sensing Image Super-resolution	This study proposes a GAN-based edge-enhancement network (EEGAN) for robust satellite image SR reconstruction and a noise-insensitive adversarial learning technique. This study showed that combining the recovered intermediate image and improved edges produces results with excellent credibility and clear content. Extensive trials on Open-Source Data sets revealed that the reconstruction performance was superior to state-of-the-art SR techniques.

Overall, the resolution of a seismic image is a complex combination of these and other factors and can vary significantly based on the specific circumstances of the survey. One promising approach to mitigate these limitations is denoising algorithms, which can filter out unwanted noise and enhance the signal-to-noise ratio of the seismic data. For example, wavelet-based denoising methods effectively reduce noise in seismic images (Yan and Chunqin, 2008).

Through the use of advanced machine learning and deep learning algorithms, AI has the potential to increase the resolution and quality of seismic images greatly. In addition to improving image resolution and quality, AI can also help automate the interpretation of seismic images, reducing the time and cost of subsurface mapping and exploration. For example, CNNs have been used to classify and identify subsurface features in seismic images (An et al., 2021). Another approach is using super-resolution reconstruction techniques, which can enhance the resolution of the seismic image beyond the limitations of the seismic source size. These techniques typically involve multiple lower-resolution images to reconstruct a higher-resolution image and have been successfully applied to various imaging modalities (Deng et al., 2021; Sun et al., 2022).

To summarise, AI has proven effective at improving seismic image resolution and reducing the effects of noise and interference. These techniques have also been used in the medical field, with similar results in enhancing resolution and minimising noise in medical imaging. However, there are still drawbacks to using AI in these applications, such as the requirement for vast volumes of training data and the possibility of overfitting. More studies are required to improve the efficiency of AI for these applications. The use of AI in seismic imaging is still in its early stages, and further research is required to optimise and validate these methods.

3. Methodology

3.1 Regional Geological Overview of the Opunake Field

The Opunake Field is located in the Taranaki region of New Zealand, on the North Island's western side. The Taranaki Basin is a large sedimentary basin formed approximately 50–60 million years ago during the early to mid-Cenozoic Era. It is known for its active volcanoes and tectonic activity, as shown in Fig. 9.1. These structural activities have shaped the geology of the area (Stagpoole and Nicol, 2008). A complex mixture of sedimentary

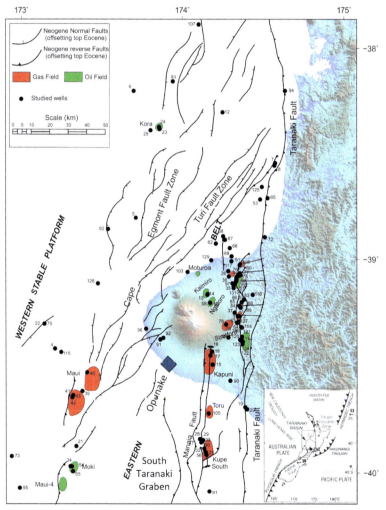

Fig. 9.1. Location map of Taranaki Basin showing simplified structural elements modified after Douglas (2005) and Rajabi et al. (2016).

(sandstone, shale, and limestone), volcanic, and metamorphic rock formations deposited by rivers and oceans at the time make up the Opunake Field's geology (Nodder, 1993).

The Opunake Field contains several layers of volcanic rock created by volcanic eruptions in the late Cenozoic Era and lies on top of the sedimentary rocks. These volcanic rocks include basalt, andesite, and rhyolite, which erupted from nearby volcanoes, including Mount Taranaki, the dominant stratovolcano in the region (Cronin et al., 2021). In addition to sedimentary and volcanic rocks, the Opunake Field also contains metamorphic rocks, formed when the sedimentary and volcanic rocks were subjected to high pressures and temperatures during tectonic activity. These metamorphic rocks include schist, gneiss, and quartzite, in which the forces of heat and pressure have altered (Nodder, 1993).

The region's tectonic activity has resulted in faulting and folding, which have an impact on the geology of the Opunake Field. The field is located on a number of anticlines and synclines, which are tectonic folds in the Earth's crust (Nicol et al., 2007). Several significant faults, including the Pohokura and Mangahewa faults, which have shaped the region's geology, border the Taranaki Basin (Stagpoole and Nicol, 2008). The Moki and Kapuni Formations are part of the Taranaki Group, a sequence of sedimentary rocks that were deposited in the Taranaki Basin during the early Cenozoic era. The main rock formations in the field are sandstones, siltstones, and mudstones of the Moki Formation, which were deposited during the Oligocene and Miocene periods (Higgs et al., 2012). The younger Kapuni Formation, comprised of sandstones, siltstones, and mudstones with interbedded coal seams, was then deposited on top of the Moki Formation (Voggenreiter, 1993). The Taranaki Group is divided into several formations based on their lithology and age.

3.2 Local Geological Overview of the Opunake Field

The Opunake Field is an oil and gas field located in the Kapuni Formation, the uppermost formation in the Taranaki Group. The field primarily comprises sandstones and siltstones, the primary reservoirs for oil and gas. The field is also rich in coal, mined and used for energy production (McBeath, 1977).

The field is located in an area with high tectonic activity and has experienced several earthquakes and volcanic eruptions in the past. These events have significantly impacted the geology of the field, including the folding and faulting of the rock formations and the alteration of the sedimentary layers. The geological history of Opunake Field can be divided into three main periods: the Paleozoic, the Mesozoic, and the Cenozoic.

During the Paleozoic period, the area was part of a shallow sea that covered much of New Zealand. The sea was home to a variety of marine life, and over time, the remains of these organisms were buried and compacted into sedimentary rocks. These rocks, which include sandstone, shale, and limestone, are the oldest in the area and can be found at the base of the Opunake Field.

During the Mesozoic period, the area continued to experience tectonic activity, which caused the land to rise and form mountains. These mountains, now long gone, were made up of sandstone, shale, and limestone. Later, these rocks were buried in sedimentary layers made up of shale, sandstone, and coal.

During the Cenozoic period, the area experienced further tectonic activity and experienced several periods of erosion. These activities caused the land to be reshaped, exposing the sedimentary layers to the surface. The area is now home to a variety of rock formations, including sandstone, shale, and limestone, as well as coal and oil deposits.

Overall, the geological history of Opunake Field is a complex and dynamic geological system reflecting the millions of years of tectonic activity, volcanic eruptions, and sedimentary processes in the Taranaki region. These activities have resulted in a wide range of rock formations and sedimentary layers that have formed over the years. These rocks and sediments provide a rich resource for geological study and exploration. The Opunake Field is a valuable oil and gas production resource that has also played an important role in developing New Zealand's energy industry (McBeath, 1977).

This study was performed using a software environment consisting of a system unit running on a Windows 10 Pro-64-bit operating system, CUDA Toolkit 10.2, and the Pytorch 2.0 framework. The experimental hardware environment of the unit is the next-generation AMD Radeon Pro™ and the highest performing NVIDIA Quadro® professional graphics capable of 2-petaFLOPS tensor performance. This study was run on a single NVIDIA RTX 8000 GPU, quadruple Intel (R) Xeon (R) W-2223 CPU running at 3.60 GHz, 32.0 GB of DDR4-2666 MHz DRAM. Throughout the course of this research, the pre-trained DCNN model of the Geoplat AI software was utilised. The model was pre-trained using both synthetic and real data.

3.3 Deep Convolutional Neural Network in Seismic Image Resolution

Seismological data are obtained through the reflection of seismic wave signals, and after a series of processing steps, the seismic amplitude data are obtained and prepared for interpretation. The amount of seismic data, including 3D volume data, is constantly increasing as exploration technology advances. A network's calculation cost rises if it is directly trained using 3D data.

The Geoplat AI software was used to import the entire 3D seismic volume of the Opunake Field for this study. The software includes a pre-trained DCNN model that is based on computer vision. The model was trained on

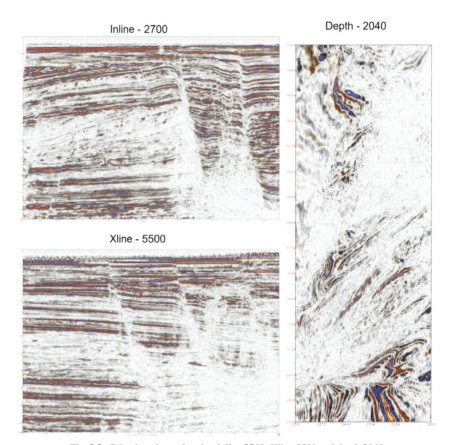

Fig. 9.2. Seismic volume showing Inline 2700, Xline 5500 and depth 2040.

a large and varied 3D synthetic seismic dataset image library that considers various complex geological factors and imaging quality. The model's accuracy was verified and successfully applied to the Opunake Field to generate a conditioned probability volume. The widely known ResU-net architectural framework, which was pre-trained, was used to segment the 3D seismic images obtained from the Opunake Field. A signal gain was added to amplify and restore regions with amplitude expression coefficient loss. This additional gain aided in the processing of low-quality seismic data. Figure 9.2 displays the inline, crossline, and depth sections of the Opunake Field.

3.3.1 Simplified Architecture of Residual U-net

The Simplified Architecture of Residual U-Net (ResU-Net) is a convolutional neural network designed for image segmentation tasks. It is an extension of the U-Net architecture, known for its ability to perform well on various image segmentation tasks. The ResU-Net architecture consists of an encoder and a decoder, connected by a series of skip connections. The encoder is responsible for extracting features from the input image, while the decoder is responsible for reconstructing the image from the features. The skip connections allow the decoder to access features from multiple scales, which helps improve the model's accuracy.

One key difference between the ResU-Net and U-Net architecture is using residual blocks in the encoder and decoder. A residual block is a block of layers that learns the residual mapping between the input and the output rather than the actual mapping. This residual block helps reduce the number of parameters in the model and improves its learning ability. The mathematical formulation of the ResU-Net architecture can be described as follows:

Let X be the input image and Y be the output image (i.e., the segmentation map). The encoder consists of a series of convolutional layers and residual blocks, which can be written as:

$$E(X) = F(X) + X \qquad (9.1)$$

where $F(X)$ is the residual mapping learned by the residual block.

The decoder consists of a series of upsampling layers and residual blocks, which can be written as:

$$D(E(X)) = G(E(X)) + E(X) \qquad (9.2)$$

where $G(E(X))$ is the residual mapping learned by the residual block.

The model's final prediction is obtained by applying a convolutional layer to the decoder output:

$$Y' = Conv(D(E(X))) \qquad (9.3)$$

A schematic framework of the ResU-Net architecture is shown in Fig. 9.3 below.

In the simplified schematic framework of a ResNet, the input image is passed through a series of convolutional layers and residual blocks. The convolutional layers extract features from the input image, while the residual blocks are responsible for learning the residual mapping between the input and the output. The output of the residual block is then added to the original input to produce the final output of the block. The output of the final residual block is passed through a series of upsampling layers, increasing the output's resolution. This upsampled output is then passed through a final convolutional layer, which produces the final prediction of the model. Overall, the residual blocks and upsampling layers form a U-shaped structure, with the convolutional layers forming the encoder and the upsampling layers forming the decoder. The skip connections between the encoder and decoder allow

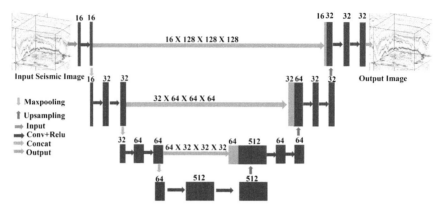

Fig. 9.3. A simplified end-to-end U-network architecture.

the model to access features from multiple scales, which helps improve the model's accuracy.

The key difference between a standard CNN and a residual neural network lies in how information flows through the network. In a standard CNN, the layers learn the actual mapping between the input and the output. In a residual neural network, the layers learn the residual mapping between the input and the output, which is then added to the original input to produce the final output. This residual network helps reduce the number of parameters in the model and improves its learning ability. Overall, the ResU-Net architecture combines the ability of the U-Net to capture multi-scale features with the ability of residual blocks to reduce the number of parameters and improve learning. This technique augments the convolutional neural network architecture and makes it a powerful tool for image resolution tasks.

3.4 Training and Testing Process

The methodology for training and testing the two CNN models in this study can be broken down into several steps:

1. *Data preparation*: The first step was to collect and prepare the dataset that would be used to train and test the models. A dataset of low-quality seismic volumes and their corresponding high-quality versions from the Opunake field, which were affected by various imaging issues such as noise and resolution degradation, was used in this research. The dataset was divided into two sections: training and testing. The models were trained using the training set, and their performance was evaluated using the test set.

2. *Model building*: The conventional CNN model was constructed using a standard architecture with several convolutional and pooling layers,

followed by a fully connected layer. The GAN model, on the other hand, was built using a generator and a discriminator network. The generator network was used to generate new high-quality seismic volumes. In contrast, the discriminator network was used to differentiate between the generated and the original high-quality seismic volumes.

3. *Model training*: The models were trained using the training set. The conventional CNN model was trained to improve low-quality seismic volumes by lowering noise and increasing resolution. The GAN model was trained to generate new high-quality seismic volumes that were similar to the original high-quality seismic volumes.

4. *Model evaluation*: Once the models were trained, they were evaluated using the test set. The performance of the models was evaluated using three commonly used image quality metrics: PSNR, SSIM, and SNR. PSNR measures the ratio between the maximum possible power of a signal and the power of the noise that corrupts the signal. SSIM measures the structural similarity between two images, while SNR measures the strength of a signal relative to the amount of noise present in the signal.

5. *Model comparison*: The performance of the models was compared based on the results of the evaluation. The PSNR, SSIM, and SNR values for each model were calculated for the test set and compared to determine which model performed better.

6. *Model selection*: Based on the comparison of the two models, the one that performed better regarding the image quality metrics was selected as the model of choice.

It is important to note that this process of training and testing models using a dataset is an iterative process. The model architecture, parameters and dataset itself can be fine-tuned and optimised. Additionally, visual inspection of the enhanced images can also be used to evaluate the results.

3.5 Criteria for Model Evaluation

In estimating the performance of the DCNN image resolution, the models' errors were assessed using the following metrics.

1. SNR measures the strength of a signal concerning the volume of total noise present in the signal. In image processing, SNR is generally used to measure the quality of an image. SNR is a simple metric and provides a single number that can be used to compare the quality of an image. However, it cannot give a comprehensive evaluation of an image's quality as it only considers the values and does not take into account the visual perception. It is common to use other metrics in conjunction with

SNR to get a complete evaluation of image quality, such as the SSIM and PSNR, which are more robust and take into account both luminance and structural information. The formula for SNR is:

$$SNR = 10 \times \log_{10}(Peak\ signal\ power/Mean\ square\ error) \quad (9.4)$$

2. PSNR measures the quality of a reconstructed image compared to a reference image. It is defined as the ratio of the peak signal power to the power of the noise in the reconstructed image. This metric is mathematically expressed as;

$$PSNR = 10 \times \log_{10}(MAX^2/MSE) \quad (9.5)$$

where MAX is the maximum possible pixel value of the image, and MSE is the mean squared error between the conditioned and original volume.

3. SSIM is a measure of the similarity between two images. It considers the luminance, contrast, structure, mean, standard deviation and cross-covariance of the image pixels and gives an understanding of the structural similarity between two images. SSIM is mathematically written as;

$$SSIM(x,y) = \frac{(2 \times \mu_x \times \mu_y + C1) \times (2 \times \Sigma_{xy} + C2)}{((\mu_{x^2} \times \mu_{y^2} + C1) \times (\Sigma_{x^2} + \Sigma_{y^2} + C2))} \quad (9.6)$$

where μ_x and μ_y are the pixel values' means of images x and y, respectively, Σ_x and Σ_y are the pixel values' standard deviations of images x and y, respectively, Σ_{xy} is the pixel values' covariance of images x and y and C_1 and C_2 are constants.

4. Results and Discussion

4.1 Conditioned Seismic Volume

In Figs. 9.4, 9.5, and 9.6, you will note the high level of improvement in the conditioned image resolution volumes for inlines, crosslines, and depth, respectively. In this study, we employed two CNN models, named the mean resolution volume and super-resolution volume based on their performance, to increase the quality of seismic images. The results showed massive improvements visually, especially in the deeper parts with low quality. Specifically, the super-resolution volume model achieved visually better results. While the use of deep learning in seismic image processing has demonstrated significant improvements in image quality, it is important to note that the models are not without limitations. One of the key limitations is the presence of false geological structures, which can be introduced during the

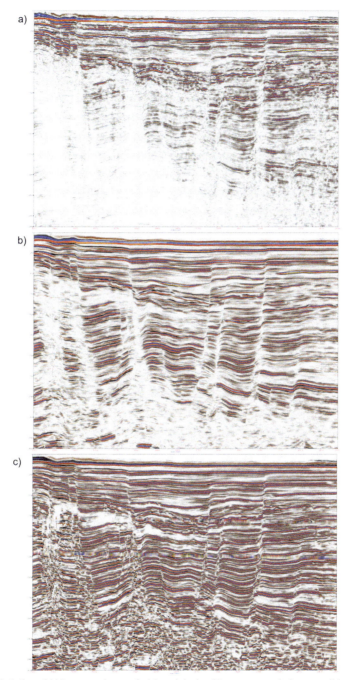

Fig. 9.4. Inline 2000 comparison of (a) original, (b) mean resolution conditioned, and (c) super-resolution conditioned volumes.

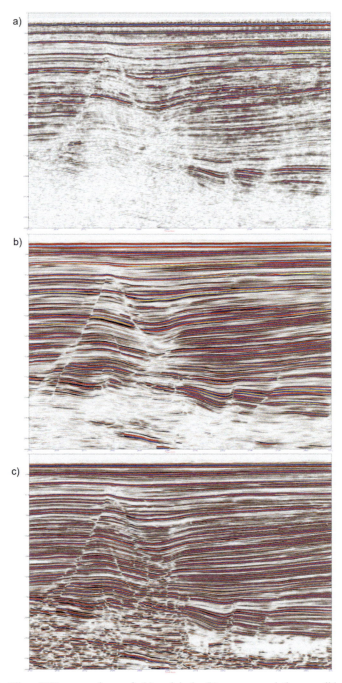

Fig. 9.5. Xline 3000 comparison of (a) original, (b) mean resolution conditioned, and (c) super-resolution conditioned volumes.

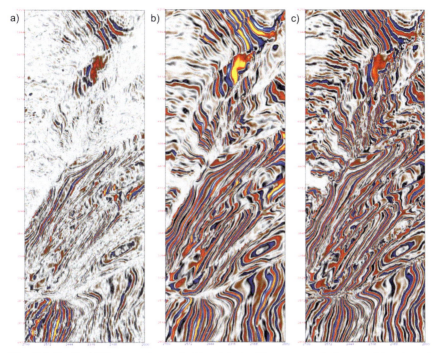

Fig. 9.6. Depth 1400 comparison of (a) original, (b) mean resolution conditioned, and (c) super-resolution conditioned volumes.

image enhancement process. Additionally, the process of enhancing the image can lead to the loss of some important geological features, which can impact the accuracy of subsequent interpretations.

The success of these models can be attributed to their ability to leverage the power of deep learning and computer vision in geophysics and seismic image processing. The models were able to extract more information from the input data, which led to significant improvements in image quality. Further investigation using PSNR and SSIM will be employed to critically evaluate the models' performance. PSNR is generally more commonly used and can provide good results if the images have a similar dynamic range and colour balance. Meanwhile, SSIM is considered a better metric of perceived image quality. This metric will enable us to accurately assess which model is better in terms of image quality improvement.

4.2 Model Evaluation

In discussing the results of image quality assessment metrics such as PSNR, SSIM, and SNR, it is important to consider the context of the images being compared and the specific application for which they will be used. It is important to understand that these metrics each provide a different perspective on the quality of the enhanced or denoised image.

PSNR is a simple metric that measures the ratio of the maximum possible power of a signal to the power of the noise that corrupts the signal. It is generally used to compare the original image to the processed image in order to assess the quality of image enhancement and denoising methods. A higher PSNR value indicates a higher quality image with less noise. From the results in Fig. 9.7, the mean resolution model achieved the highest PSNR compared to the super-resolution model. This metric is easy to understand and compute. However, it cannot give a comprehensive evaluation of an image's quality as it only considers the values and does not take into account the visual perception.

SSIM takes into account the mean, standard deviation, and cross-covariance of the image pixels and gives an understanding of the structural similarity between two images. Based on the data it takes into account when computing, SSIM is a more reliable metric for evaluating image quality than PSNR, but it requires more computation than PSNR. This metric is based on the idea that the human visual system is highly adapted to recognise small changes in structural information, such as edges and textures, rather than small changes in pixel intensities. It is typically used as a measure of image quality in signal processing applications. However, it cannot give a comprehensive evaluation of an image's quality as it only considers the values and does not take into account the visual perception. As such, it is imperative to combine it with other metrics like the PSNR. A higher SSIM value indicates a higher quality image with less noise. As such, the mean resolution model achieved the highest SSIM compared to the super-resolution model, which is shown in Fig. 9.8. The SSIM is usually measured between 0 and 1. However, for this study, the original seismic volume was used as a standard for comparison. As such, the initial SSIM was 1, whiles the improvements gained by the models were estimated between 0 and 1 to capture the increased SSIM. The mean resolution model achieved an improved SSIM of 0.4, whiles the super-resolution resulted in an SSIM of 0.31. This result shows that not all high resolution images should be classified as the best model. Some enhancements may result in the loss of important geological structures through false improvements, as typified by the SSIM results.

When evaluating image enhancement or denoising methods, PSNR, SSIM, and SNR can be used together to provide a more comprehensive image quality evaluation. For example, a method that provides a high PSNR value and a high SSIM value is likely to produce an image that is both visually vivid and accurate in terms of preserving the original information. Additionally, a high SNR value means that the signal is stronger than the noise, and it indicates that the denoising method performed well in removing the noise. It is also important to note that even though the PSNR and SSIM values may

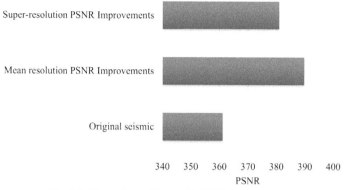

Fig. 9.7. Comparison of the models PSNR improvements.

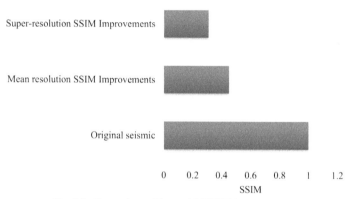

Fig. 9.8. Comparison of the models' SSIM improvements.

be high, the visual perception of the image can be different. That is why it is good practice to use multiple metrics to get a more comprehensive evaluation and to consider the specific application for which the image will be used. Hence, when there is the need to make an image resolution very high, the super-resolution model performs better, especially in the shallower part of the seismic volume. However, when the model is needed to enhance the seismic volume for interpretation, the mean resolution model is the better of the 2 due to the fact that more geological structures and features were preserved while exhibiting fewer falsely enhanced areas or less confident resolution in the deeper parts of the subsurface.

Figure 9.9 presents the spectral analysis results performed on the original, mean, and super-resolution volumes along inline 2000. Our findings indicate that the original volumes had an SNR of 70.9, which was significantly improved upon by the mean resolution volume achieving an SNR of 212.5. Interestingly, despite the improvement in spatial resolution, the

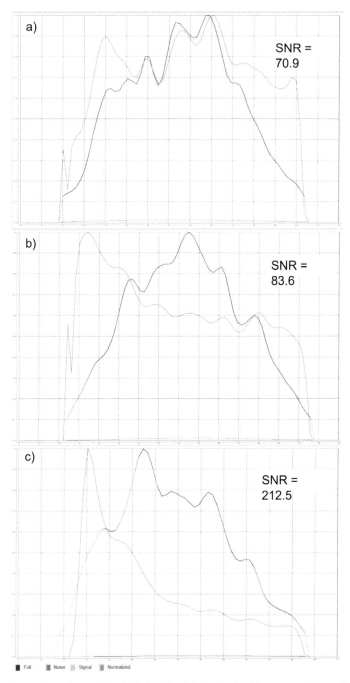

Fig. 9.9. Comparison of spectral analysis for (a) original seismic, (b) super-resolution volume, and (c) mean resolution volume.

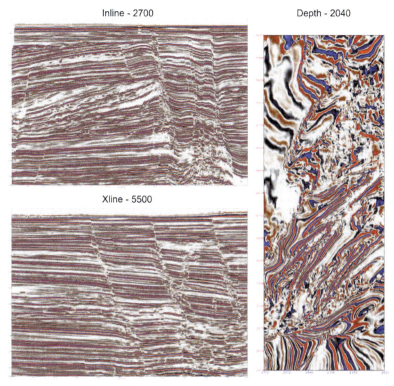

Fig. 9.10. Super-resolution conditioned volume showing Inline 2700, crossline 5500, and depth 2040.

super-resolution volume only recorded an SNR of 83.6, suggesting that while super-resolution techniques can improve the resolution of seismic data, they may not always result in better SNR. Based on the spectral analysis results, the mean volume outperformed the super-resolution volume due to its ability to preserve low-frequency information, leading to a better SNR. Additionally, the super-resolution volume amplifies the noise in the data, resulting in a lower SNR. This is shown in Figs. 9.10 and 9.11 representing the pictorial depiction of the Super-resolution conditioned volume and the Mean resolution conditioned volume respectively.

Our study highlights the potential of using the mean resolution (GAN) techniques to enhance the seismic data quality and emphasises the importance of evaluating the effectiveness of super-resolution (CNN) techniques on a case-by-case basis. Overall, our findings provide valuable insights for the seismic imaging community in their efforts to enhance the quality of subsurface images for accurate interpretation and analysis.

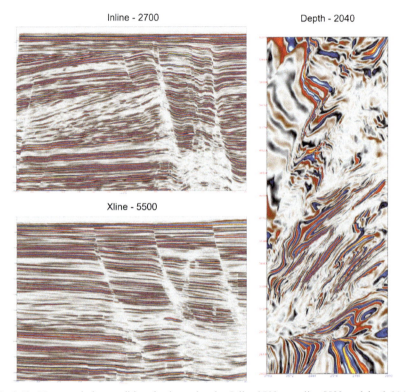

Fig. 9.11. Mean resolution conditioned volume showing Inline 2700, crossline 5500, and depth 2040.

5. Conclusions

Insufficient labelled samples have hindered the progress of deep learning application in seismic interpretation. In this study, we aimed to enhance an image using AI by building and comparing two different CNN models. The first model was a traditional CNN, while the second model was a more recent architecture known as a GAN.

The conventional CNN (super-resolution) model was trained using a dataset of low-quality images and their corresponding high-quality versions. The model was then used to enhance the test images by increasing their resolution and reducing noise. On the other hand, the GAN (mean resolution) model was trained using a dataset of high-quality images. The generator network of the GAN was used to generate new high-quality images, which were then compared with the original test images.

The performance of both models was evaluated using three commonly used image quality metrics: PSNR, SSIM, and spectral analysis. PSNR measures the ratio between the maximum possible power of a signal and the power of the noise that corrupts the signal. SSIM measures the structural similarity between two images, while SNR measures the strength of a signal relative to the amount of noise present in the signal.

The evaluation results showed that both models could enhance the images to a certain extent. However, the GAN model performed better than the conventional CNN model in terms of all three metrics. The GAN model achieved a higher PSNR, SSIM, and SNR value than the conventional CNN model, indicating that the generated images were of higher quality and contained less noise.

In conclusion, the study successfully demonstrated the use of AI for image enhancement using CNN models. Both traditional CNN and GAN models were able to enhance the images. However, the GAN model performed better in terms of all three image quality metrics. The GAN model showed superior performance in terms of PSNR, SSIM, and SNR, which means it was able to generate images with less noise and more structural similarity. Based on these results, it can be concluded that the GAN model is a better choice for image enhancement tasks. The proposed methodology aims to encourage the utilisation and exploration of deep learning techniques in seismic interpretation.

Data and Software Availability

The data used in this study is available by request through the corresponding author. To access the software used in this article, contact the developers at https://geoplat.ai/#demo.

Acknowledgement

The authors thank the University Teknologi Petronas and the Centre for Subsurface Seismic Imaging for supporting this work and to Geoplat AI for providing the software for this work.

References

Aminzadeh, F. and Dasgupta, S.N. 2013. *Fundamentals of Petroleum Geophysics*. pp. 37–92. https://doi.org/10.1016/B978-0-444-50662-7.00003-2.

An, Y., Guo, J., Ye, Q., Childs, C., Walsh, J. and Dong, R. 2021. Deep convolutional neural network for automatic fault recognition from 3D seismic datasets. *Comput. Geosci.* 153: 104776. https://doi.org/10.1016/j.cageo.2021.104776.

Cengiz, E., Kelek, M.M., Oğuz, Y. and Yılmaz, C. 2022. Classification of breast cancer with deep learning from noisy images using wavelet transform. *Biomedical Engineering/ Biomedizinische Technik* 67: 143–150. https://doi.org/10.1515/bmt-2021-0163.

Chen, Y., Xie, Y., Zhou, Z., Shi, F., Christodoulou, A.G. and Li, D. 2018. Brain MRI super resolution using 3D deep densely connected neural networks. *2018 IEEE 15th International Symposium on Biomedical Imaging (ISBI 2018)* April, pp. 739–742. https://doi.org/10.1109/ISBI.2018.8363679.

Cronin, S.J., Zernack, A.v., Ukstins, I.A., Turner, M.B., Torres-Orozco, R., Stewart, R.B., Smith, I.E.M., Procter, J.N., Price, R., Platz, T., Petterson, M., Neall, V.E., McDonald, G.S., Lerner, G.A., Damaschcke, M. and Bebbington, M.S. 2021. The geological history and hazards of a long-lived stratovolcano, Mt. Taranaki, New Zealand. *New Zealand Journal of Geology and Geophysics* 64: 456–478. https://doi.org/10.1080/00288306.2021.1895231.

Deng, M.-D., Jia, R.-S., Sun, H.-M. and Zhang, X.-L. 2021. Super-resolution reconstruction of seismic section image via multi-scale convolution neural network. *E3S Web of Conferences* 303: 01058. https://doi.org/10.1051/e3sconf/202130301058.

Douglas, A. 2005. Slow slip on the northern Hikurangi subduction interface, New Zealand. *Geophys. Res. Lett.* 32: L16305. https://doi.org/10.1029/2005GL023607.

Farfour, M., Gaci, S., El-Ghali, M. and Mostafa, M. 2021. A review about recent seismic techniques in shale-gas exploration. pp. 65–80. *In*: *Methods and Applications in Petroleum and Mineral Exploration and Engineering Geology*. Elsevier. https://doi.org/10.1016/B978-0-323-85617-1.00012-6.

Gondara, L. 2016. Medical image denoising using convolutional denoising autoencoders. pp. 241–246. *In*: *2016 IEEE 16th International Conference on Data Mining Workshops (ICDMW)*. IEEE. https://doi.org/10.1109/ICDMW.2016.0041.

Gray, S.H. 2014. Seismic imaging. pp. S1-1-S1-16. *In*: *Encyclopedia of Exploration Geophysics. Society of Exploration Geophysicists*. https://doi.org/10.1190/1.9781560803027.entry4.

Haldar, S.K. 2018. Exploration geophysics. pp. 103–122. *In*: *Mineral Exploration*. Elsevier. https://doi.org/10.1016/B978-0-12-814022-2.00006-X.

Higgs, K.E., King, P.R., Raine, J.I., Sykes, R., Browne, G.H., Crouch, E.M. and Baur, J.R. 2012. Sequence stratigraphy and controls on reservoir sandstone distribution in an Eocene marginal marine-coastal plain fairway, Taranaki Basin, New Zealand. *Mar. Pet. Geol.* 32: 110–137. https://doi.org/10.1016/J.MARPETGEO.2011.12.001.

Jiang, K., Wang, Z., Yi, P., Wang, G., Lu, T. and Jiang, J. 2019. Edge-enhanced GAN for remote sensing image super-resolution. *IEEE Transactions on Geoscience and Remote Sensing* 57: 5799–5812. https://doi.org/10.1109/TGRS.2019.2902431.

Jo, Y., Choi, Y., Seol, S.J. and Byun, J. 2022. Machine learning-based vertical resolution enhancement considering the seismic attenuation. *J. Pet. Sci. Eng.* 208: 109657. https://doi.org/10.1016/j.petrol.2021.109657.

Ker, S., Marsset, B., Garziglia, S., le Gonidec, Y., Gibert, D., Voisset, M. and Adamy, J. 2010. High-resolution seismic imaging in deep sea from a joint deep-towed/OBH reflection experiment: Application to a mass transport complex offshore Nigeria. *Geophys. J. Int.* 182: 1524–1542. https://doi.org/10.1111/j.1365-246X.2010.04700.x.

Lan, T., Zeng, Z., Han, L. and Zeng, J. 2023. Seismic data denoising based on wavelet transform and the residual neural network. *Applied Sciences* 13(1): 655. https://doi.org/10.3390/app13010655.

Li, J., Wu, X. and Hu, Z. 2022. Deep learning for simultaneous seismic image super-resolution and denoising. *IEEE Transactions on Geoscience and Remote Sensing* 60. https://doi.org/10.1109/TGRS.2021.3057857.

McBeath, D.M. 1977. Gas-condensate fields of the Taranaki basin, New Zealand. *New Zealand Journal of Geology and Geophysics* 20: 99–127. https://doi.org/10.1080/00288306.1977.10431594.

Mukherjee, S., Bell, R.S., Barkhouse, W.N., Adavani, S., Lelièvre, P.G. and Farquharson, C.G. 2022. High-resolution imaging of subsurface infrastructure using deep learning artificial intelligence on drone magnetometry. *The Leading Edge* 41: 462–471. https://doi.org/10.1190/TLE41070462.1.

Nicol, A., Mazengarb, C., Chanier, F., Rait, G., Uruski, C. and Wallace, L. 2007. Tectonic evolution of the active Hikurangi subduction margin, New Zealand, since the Oligocene. *Tectonics* 26(4): 1–24. https://doi.org/10.1029/2006TC002090.

Nodder, S.D. 1993. Neotectonics of the offshore Cape Egmont Fault Zone, Taranaki Basin, New Zealand. *New Zealand Journal of Geology and Geophysics* 36: 167–184. https://doi.org/10.1080/00288306.1993.9514566.

Orozco-del-Castillo, M.G., Ortiz-Alemán, C., Urrutia-Fucugauchi, J., Martin, R., Rodriguez-Castellanos, A. and Villaseñor-Rojas, P.E. 2014. A genetic algorithm for filter design to enhance features in seismic images. *Geophys. Prospect.* 62: 210–222. https://doi.org/10.1111/1365-2478.12026.

Otchere, D.A., Abdalla Ayoub Mohammed, M., Ganat, T.O.A., Gholami, R. and Aljunid Merican, Z.M. 2022a. A novel empirical and deep ensemble super learning approach in predicting reservoir wettability via well logs. *Applied Sciences* 12: 2942. https://doi.org/10.3390/app12062942.

Otchere, D.A., Tackie-Otoo, B.N., Mohammad, M.A.A., Ganat, T.O.A., Kuvakin, N., Miftakhov, R., Efremov, I. and Bazanov, A. 2022b. Improving seismic fault mapping through data conditioning using a pre-trained deep convolutional neural network: A case study on Groningen field. *J. Pet. Sci. Eng.* 213: 110411. https://doi.org/10.1016/J.PETROL.2022.110411.

Picetti, F., Lipari, V., Bestagini, P. and Tubaro, S. 2019. Seismic image processing through the generative adversarial network. *Interpretation* 7: SF15–SF26. https://doi.org/10.1190/INT-2018-0232.1.

Rajabi, M., Ziegler, M., Tingay, M., Heidbach, O. and Reynolds, S. 2016. Contemporary tectonic stress pattern of the Taranaki Basin, New Zealand. *J. Geophys. Res. Solid Earth* 121: 6053–6070. https://doi.org/10.1002/2016JB013178.

Roy Chowdhury, K. 2011. *Seismic Data Acquisition and Processing*. Dordrecht: Springer, pp. 1081–1097. https://doi.org/10.1007/978-90-481-8702-7_52.

Selley, R.C. and Sonnenberg, S.A. 2015. Methods of exploration. pp. 41–152. *In: Elements of Petroleum Geology*. Elsevier. https://doi.org/10.1016/B978-0-12-386031-6.00003-5.

Shi, F., Cai, N., Gu, Y., Hu, D., Ma, Y., Chen, Y. and Chen, X. 2019. DeSpecNet: A CNN-based method for speckle reduction in retinal optical coherence tomography images. *Phys. Med. Biol.* 64: 175010. https://doi.org/10.1088/1361-6560/AB3556.

Stagpoole, V. and Nicol, A. 2008. Regional structure and kinematic history of a large subduction back thrust: Taranaki Fault, New Zealand. *J. Geophys. Res. Solid Earth* 113. https://doi.org/10.1029/2007JB005170.

Sun, Q.F., Xu, J.Y., Zhang, H.X., Duan, Y.X. and Sun, Y.K. 2022. Random noise suppression and super-resolution reconstruction algorithm of seismic profile based on GAN. *J. Pet. Explor. Prod. Technol.* 12: 2107–2119. https://doi.org/10.1007/S13202-021-01447-0/FIGURES/7.

Talwani, M. and Kessinger, W. 2003. Exploration geophysics. pp. 709–726. *In: Encyclopedia of Physical Science and Technology*. Elsevier. https://doi.org/10.1016/B0-12-227410-5/00238-6.

Voggenreiter, W.R. 1993. Structure and evolution of the Kapuni Anticline, Taranaki Basin, New Zealand: Evidence from the Kapuni 3D seismic survey. *New Zealand Journal of Geology and Geophysics* 36: 77–94. https://doi.org/10.1080/00288306.1993.9514556.

Wang, E. and Nealon, J. 2019. Applying machine learning to 3D seismic image denoising and enhancement. *Interpretation* 7: SE131–SE139. https://doi.org/10.1190/INT-2018-0224.1.

Yan, F. and Chunqin, Z. 2008. Seismic data denoising based on second wavelet transform. pp. 186–189. *In: 2008 International Conference on Advanced Computer Theory and Engineering*. IEEE. https://doi.org/10.1109/ICACTE.2008.118.

CHAPTER 10
Petroleum Reservoir Characterisation
A Review from Empirical to Computer-Based Applications

Ebenezer Ansah,[1,]* *Anthony Ewusi*,[2] *Eric Thompson Brantson*[3]
and *Jerry S.Y. Kuma*[2]

1. Introduction

In the early days, geological studies entailed reservoir description by integrating geologic, geophysical, and well logging data (Yu et al., 2011). Given this, practical problems related to geology and engineering highlighted the importance of integrated reservoir characterisation. That led to immense research and application of various reservoir characterisations in the oil and gas industries.

According to Ma (2011), reservoir characterisation is the study of the properties of a reservoir using various field specialisations (geological, geophysical, petrophysical, and engineering), including uncertainty analysis of geological, engineering data, and spatial variations. There exists a correlation among these sparse data sources, and analysing them can bring out a great understanding of the reservoir. The most used data for reservoir characterisation are seismic data (2D, 3D, and 4D), well logs, and core data. Recent advancement has seen increased development in the application of

---

[1] Department of Petroleum Geosciences and Engineering, School of Petroleum Studies, University of Mines and Technology, Tarkwa, Ghana.
[2] Department of Geological Engineering, Faculty of Geosciences and Environmental Studies, University of Mines and Technology, Tarkwa, Ghana.
[3] Department of Petroleum and Natural Gas Engineering, School of Petroleum Studies, University of Mines and Technology, Tarkwa, Ghana.
* Corresponding author: ansahessel@gmail.com

microseismic in the field of reservoir characterisation. In the characterisation process, well logs provide a high vertical resolution, while seismic data provide a high lateral resolution since it covers a large area. The core analysis gives a precise insight into the *in-situ* behaviour of the reservoir (highest resolution), but its acquisition is expensive and needs huge analysis effort, making it limited. Thus, all the available data sources complement each other to effectively attain a complete reservoir description.

To highlight a good reservoir model, the correlation in the various data sets is insufficient since the data have some level of uncertainty, imprecision, and incompleteness (Verma et al., 2012). Therefore, Artificial Intelligence (AI) techniques have been one of the widely used modelling practices to handle such problems relating to the data set used for reservoir characterisation (Verma et al., 2012). Given the training dataset, a good model can be trained for complex problems. Literature has highlighted that AI methodologies have the strength of ameliorating models by integrating with other computing techniques (Anifowose et al., 2017; Nikravesh, 2003). These computing techniques such as genetic algorithms, and fuzzy logic help to correct some various limitations (such as overfitting, parameter selection, etc.) of the AI algorithms.

This paper provides a structured overview of the application of AI in reservoir characterisation. Several reviews highlighting the use of AI in reservoir properties can be found in (Anifowose et al., 2015; Anifowose et al., 2017; Nikravesh and Aminzadeh, 2001; Otchere et al., 2021a; Saikia et al., 2020; Yu et al., 2011). In this paper, the emphasis is on the empirical approach, AI application and its enhancement with other computational models, and recent AI advancement for reservoir property prediction. The paper will also address some challenges and perspectives on the way forward for reservoir characterisation.

The following is how the paper is structured. The second section provides an overview of the empirical correlation used for petrophysical reservoir characterisation. Section 3 highlights the application of fractals analysis in reservoir characterisation. Section 4 is the application of AI in petrophysical reservoir characterisation with an emphasis on the application of artificial neural network (ANN) and support vector machine (SVM). Section 5 looks at the review of lithology and facies analysis while Section 6 is the contribution of seismic data in reservoir property prediction. Section 7 looks at the application of hybrid AI in reservoir characterisation. Section 8 highlights the summary of the individual AI algorithms used in reservoir characterisation with key consideration of the data input, activation function, output, and the statistical measures for the AI algorithms. Section 9 addresses the challenges and the way forward for reservoir characterisation. Section 10 concludes the study and spells out the major keys drawn from the present review.

2. Empirical Models for Petrophysical Property Prediction

Reservoir study using petrophysical analysis is one of the useful tools because it provides the basic input data for integrating reservoir characterisation and reservoir resource evaluation. The empirical models for reservoir characterisation are represented by the correlational relationship between the input and output variables represented by mathematical expressions. These empirical models are vital due to their potency to correlate measurable variables to important variables which may be difficult to determine. In the light of this, Reza et al. (2017) highlighted the importance of well log (input variable) analysis and its application in earth science and relation to reservoir property prediction (output variables). They further commented on the application in the various stages of an oilfield and their variation to different methods used to determine reservoir properties. The petrophysical analysis embodies the prediction of the reservoir parameters (which include but are not limited to porosity (Phi), permeability (K), water saturation (Sw), and lithology). To date, the reliable way of determining the true reservoir properties is laboratory measurement (porosity, permeability, grain density, and saturation) on core samples and well-test interpretation. However, these core samples are only limited to small intervals due to the cost involved in acquiring them. Therefore, the determination of the reservoir properties for the well interval relies on the interpretation of the vast petrophysical well logs acquired. The interpretation requires a systematic step in processing the well log data. Due to this, the results of each step always have an impact on the subsequent step.

2.1 Porosity and Permeability Prediction Models

According to Mohaghegh et al. (1996), the traditional method for estimating reservoir parameters has adopted a linear or nonlinear relationship among individual log. They highlighted in their study that the traditional approach is limited to assumptions, constants, and heterogeneity of the reservoir but Nelson (1994) in his paper reviewed some of these problems in detail. The traditional approach of estimating Phi and K from core samples are the use of models such as the Kozeny-Carman equation (Kozeny, 1927; Carman, 1937 and 1956), Wyllie and Rose model (Wyllie and Rose, 1950), and Neutron-Density relation for Phi estimation (Gaymard and Poupon, 1968), Timur correlation for K estimation (Timur, 1968) from wireline logs. The empirical relation used to estimate Phi from log response (density and neutron log) yields a total porosity (PhiT) of the formation. This PhiT is not effective to use in reservoir characterisation due to reservoir heterogeneity and issues such as the presence of an isolated pore network. Mavko and Nur (1997) underlined the fact that the Kozeny-Carman's relation observed low Phi where there is a decrease in K with a rapid decrease in Phi. To account for this, Schwartz and Kimminau (1987) introduced a consolidated model to survey the effect of multiple accumulations of cementation on Phi. Bourbie et al. (1987), also proposed a solution by introducing a variable power on the Phi in the Kozeny-Carman equation (Eq. (10.1)).

$$K = B\phi^n d^2 \qquad (10.1)$$

where K is the permeability, ϕ is the porosity, d is the diameter, and n is the variable power introduced by Bourbie et al. (1987). The n varies from 3 and 7–8 for higher porosities and lower porosities, respectively. With all the solutions, Mavko and Nur (1997), proposed a solution by accounting for the porosity percolation threshold. Their research included the percolation threshold in the Kozeny-Carman relation (Eq. (10.2)) to accurately fit the observed permeability to a well-sorted material and also extend the scope of the model.

$$K = B(\phi - \phi_c)d^2 \qquad (10.2)$$

where ϕ_c is the minimum threshold porosity. Note, the $(\phi - \phi_c)$ replacing ϕ in the Kozeny-Carman relation caters for the percolation threshold.

2.2 Saturation Prediction Models

Various models have been developed to estimate Sw which mostly depend on the geology of the formation to be evaluated. The application of these models can lead to a different prediction of Sw. The common model used for determining Sw is Archie's equation (10.3), Simandox equation (10.4), Waxman Smits' equation (10.5), Indonesian equation (10.6) and Dual water equation (10.7) correlation. An overview of the several water saturation models is presented in Table 10.1. Archie's is the most widely used model and it is applied to clean sand and carbonates having zero shaly volume (Shedid and Saad, 2017). Archie (1941) used resistivity data to establish a model to determine the Sw model for clean sandstone and carbonate reservoirs on an assumption that water is the conductive element in the reservoir. Yet, in 2016, Mehana and El-Monier (2016) indicated that Archie's assumption is not valid in shale formation since it has a high mineral and organic content. Also, Shedid and Saad (2017) highlighted the same disadvantage of Archie's model. They highlighted that the presence of clay minerals in a formation complicates the results from Archie's and may give a misleading result.

Following this, several empirical models were built to estimate Sw in shaly and shaly sand formation. All aforementioned models are associated with various uncertainties. These uncertainties can also be dependent on the accuracy and precision of the well logs and core measurements. Note that,

Table 10.1. Characteristics of various water saturation models.

Sw model	Characteristics
Archie's model (Archie 1942)	Archie's equation (Archie 1942) shows a relationship between water saturation to the true permeable formation resistivity, the formation porosity and the formation water resistivity. The challenge, therefore, arises due to the presence of shale in the reservoir which is a conductive medium and hence is against the original assumptions of Archie's equation, which was a clean sandstone reservoir (Archie 1942). The presence of shale causes a disparity in the reading of the total resistivity of the reservoir and brings about an overshot in the water saturation predicted by Archie's equation (Archie 1942).
Simandoux (Simandoux, 1963)	The Simandoux model was developed to study the volumetric effects of reducing clay volume on the conductivity of the rock matrix and the overall saturation. The Simandoux model is applied in fine siltstone of clay-rich formation regardless of the specific distribution form of clay or clay applied in shaly sandstone. Simandoux experimented on only four synthetic samples using one type of clay (montmorillonite) with a constant value of porosity. Hence, the model leads to optimistic results when the porosity is less than 20%, and it cannot be relied on in low porosity situations. Also, the model does not show a volumetric balance between sandstone volume and the clay volumes with consideration of the lack of shale formation factor in the clay term making the correlation of clay effect in the model too large and hence reducing the amount of water saturation estimated (Sam-Marcus et al., 2018)
Waxman-Smits (Waxman-Smits 1968)	The Waxman–Smits model is based on laboratory measurements of resistivity, porosity and saturation of real rocks. The major assumptions of the Waxman–Smits model about clay formation and its properties are as follows: Clay surface conductivity is assumed to share a directly proportional relationship with the factor Qv (defined as the milli-equivalents of exchangeable clay counterions per unit volume of pore space), and the F^* term replicated in both the sandstone resistivity term and the shale resistivity term. This model served as the premise of the widely used dual water model. The Waxman–Smit's equation is often used as a standard against other methods, due to its high experimental backing, but the determination of CEC (Cation Exchange Capacity) is a time-consuming experiment and this is the major limitation of the Waxman–Smits model.
Indonesian (Poupon and Leveaux, 1971)	The Indonesian equation relics as a benchmark for field-based models that work reliably with log-based analysis regardless of special core analysis data. The Indonesian equation also does not particularly assume any specific shale distribution. The Indonesian model also has an extra feature as the only model considered the saturation exponent (n). According to Shedid and Saad (2017), the results from the Indonesian predictor have been obtained with a simpler equation, which is more convenient for quick interpretation.
Dual water (Clavier et al. 1977)	Dual water is an improved form of Waxman-Smits which contains irreducible water saturation and free water. In this method, it is proposed that the contribution of clay minerals to the resistivity of reservoir rock is caused by the presence of free water within the pore spaces and the bound water within the clay matrix. Dual water was developed to account for the conductivity at the surface of a clay mineral within the volume of shaly sandstone.

these uncertainties have direct implications on the economics of any project but often petrophysicists do not quantify them. To assess the implication, Bower and Fitz (2000) quantified the sensitivity associated with the dual water model to determine the most cost-effective way to reduce ambiguity and deliver a framework for good estimation. Figure 10.1 provides a flowchart of the influence of petrophysical properties on the Net Present Value (NPV) for a project. Also, a balance in capillary pressure and gravitational forces plays a vital role in the distribution of Sw above a Free Water Level (FWL) (Lian et al., 2016). Therefore, the availability of capillary pressure data from Special Core Analysis (SCAL) can help estimate the Sw at any given height (Brantson, 2022; Harrison and Jing, 2001; Olakunle et al., 2015; Rudyk and Al-Lamki, 2015). Possibly, a good saturation height model can be used to estimate the Sw value away from the cored well location provided that the geology and sedimentological setting of the predicted environment is known (Hu and Chen, 2012; Lian et al., 2016; Luo et al., 2013).

$$Sw = \sqrt[n]{\frac{(a*Rw)}{Rt*Phie^m}} \qquad (10.3)$$

$$\frac{1}{Rt} = \frac{\phi^m * Sw^n}{a*Rw} + \frac{Vsh*Sw}{Rsh} \qquad (10.4)$$

$$\frac{1}{Rt} = \frac{\phi_T^{m*} * SwT^n}{a*Rw} * \left(1 + B*Qv \frac{Rw}{SwT}\right) \qquad (10.5)$$

$$\frac{1}{\sqrt{Rt}} = \left(\sqrt{\frac{\phi^m}{a*Rw}} + \frac{Vcl^{(1-(Vcl/2))}}{\sqrt{Rcl}}\right) * Sw^{n/2} \qquad (10.6)$$

$$\frac{1}{Rt} = \frac{\phi_T^{m*} * SwT^n}{a} * \left(\frac{1}{Rw} + \frac{Swb}{SwT}\left(\frac{1}{Rwb} - \frac{1}{Rw}\right)\right) \qquad (10.7)$$

where m is the cement factor, $m*$ presents the cement factor used for Dual water and Waxman-Smiths, n is saturation exponent, a represents Tortuosity factor, Vsh is wet shale volume, Sw represents effective water saturation, SwT is Total water saturation, Swb is Bounded water saturation, Rw represents Formation water resistivity, Rwb also represents Bound water resistivity, Rt is Input resistivity curve, Qv represents cation exchange capacity per unit total pore volume, B is Equivalent conductance of clay cations and T representing Formation temperature in degree centigrade.

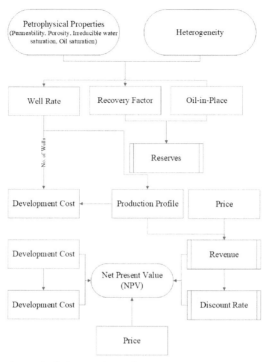

Fig. 10.1. Influent of petrophysical properties on the net present value (NPV). Modified after Bowers and Fitz (2000).

3. Fractal Analysis in Reservoir Characterisation

Many numerical fashions have been used to acquire reservoir properties which have facilitated information on the relationship between reservoir properties and the components of reservoir decision programs. Fractals have emerged as one of the effective tools for quantifying complicated reservoir structures and are extensively used to model pore shape and flow phenomena in conventional and unconventional reservoirs (Bian et al., 2020). Hewett (1986) initially developed thoughts of fractal geometry to reservoir descriptions and evaluations. Hewett and Bahrens (1990) confirmed how the distribution of reservoir properties may be regarded as a consciousness of a random function by advocating two uses of condition simulation. They utilised the flow process and properties of various scales in the reservoir and assessed the uncertainties associated with the final results using incomplete data sampling.

Sahimi (1993) handled the maximum essential elements of flow in porous rock, from modelling to fractals, percolation and current developments. Heiba et al. (1992) modelled the percolation theory of two-phase relative permeability, whilst Katz and Thompson confirmed that the porous rock interface of numerous sandstones is fractal. Shook et al. (1992), upon defining scaling as the extrapolation of the outcome of one scale to another scale, Li and Lake (1995) implemented the concept to develop a scaling procedure for a heterogeneous rock. Costa (2006) developed a fractal predictive model for permeability by incorporating the concept of fractal geometry into the fundamental Kozeny-Carman equation. The developed model was then used to predict the permeability of a variety of porous materials and discovered that the version was more adaptable to experimental results than other models. Zheng and Li (2015) developed a fractal permeability model for sandstone reservoirs that is entirely based on a fractal capillary model and Darcy's law. The version is made up of a fractal Kozeny-Carman constant that represents the microstructural parameters of the pores. Yu and Cheng (2002) created the fractal permeability model for heterogeneous porous media and delivered the set of rules for the fractal size of the pore dimension and the fractal size of tortuosity primarily based on the box-counting method. Bian et al. (2020), analyse how microporous systems affect the permeability of clay silt reservoirs according to the relationship between permeability and pore structure parameters. In addition, Dong (2020) advanced a predictive permeability model for low-permeability sandstone that takes into consideration the microstructural parameters and tortuosity. Upon comparing the two predicted model outputs to the core measured permeability, a relative average error of 8% was obtained. The reviewed literature has proven that fractal theory has been validated to be another good prospect for enhancing reservoir characterisation and can also be used to assess field-wide project performance.

4. Application of Artificial Intelligence in Petrophysical Property Prediction

In recent years, the use of technology has grown widely and the application of such technology has helped solve problems in the petroleum sector. Computer-based intelligence like ANNs and Machine Learning (ML) has helped solve petroleum-related problems such as the nonlinearity problems (Ouenes, 2000; Kaydani et al., 2012) associated with the traditional correlations. It has higher capabilities for recognising the irregularities in the multiresolution of the individual variables that are not tended to by conventional relationships. Also, in using computer-based intelligence, there are no required assumptions to be made (Otchere et al., 2021a), and its operation is based on numerical models which can identify irregularities and help lessen noise present in the dataset (Saikia et al., 2020). Olson (1998), highlighted in his study that, the application of computer-based intelligence for formation evaluation is important to establish a reliable linear relationship that exists between the estimated reservoir elastic properties from the wireline logs and corresponding sampled fluid information. The application of computer-based intelligence provides a different perspective on the implementation of well logs and also brings to light more details contained in these input logs (e.g., the relationship between Gamma Ray (GR), Deep resistivity (ILD), and Density (RHOB) with formation K). In this section, AI techniques such as ANN, SVM and fuzzy logic (FL) will be discussed by providing the main theoretical framework and some applications in reservoir characterisation.

The application of ML algorithms has been applied to several quantitative analyses of well logs for the estimation of reservoir properties (Helle et al., 2001; Huang et al., 1996; Huang and Williamson, 1997). This learning approach has proved to be simple and has provided an accurate solution for the valuation of reservoir properties using several well logs. Nikravesh et al. (2003) noted that the computational process for the estimation of the reservoir parameters seems to be more reliable because they are independent of the uncertainties which come with the logging process.

For the various AI concept and theories, Otchere et al. (2021a) have presented an in-depth report on the mathematical structure of some of the AI algorithms such as ANN, SVM, and Relevance Vector Machine (RVM). Likewise, Saika et al. (2020) and Otchere et al. (2021a) provides a detailed overview of the structure of AI algorithms and their advantages and limitations.

4.1 Artificial Neural Networks (ANNs)

In the early 1940s, McCulloch and Pitts (1943) idealised a mathematical model as a representation of the human biological neural system. This idea of the mathematical model is termed the ANN. The neurons of the human body are dedicated to information processing. These neurons are made up of cell body branches tree-like axons (which are signal transmitters) and dendrites (which are signal receivers). The cell body also includes a nucleus which produces chemicals required for the neuron. The chemical produced by the nucleus is then transferred between a neuron and its adjacent target cell through the synaptic connections found at the end of the axon branch (which are termed axon terminals). In the ANN architecture, the input signals $(x_1, x_2, \ldots, x_n)$; which serves as the chemical product or the electrical signal) are transmitted along the axons to the synaptic connections. At this point, weights $(w_1, w_2, \ldots, w_n)$ will be applied to the signals from the axon. These weights control the direction and influence of the cells. The product of the weight and signal $(w_1x_1, w_2x_2, \ldots, w_nx_n)$ will be transferred from the dendrites to the neuron cell. At the neuron cell, a summation of all the products of the weights and signals from the dendrites is made. A signal is then fired along the axon if the summation value exceeds a threshold. During this stage, an activation function (f) performs a transformation on the signal received from the axon and influences the frequency of the final signal. Some common activation functions employed in the design of an ANN architecture are sigmoid, gaussian, tangent hyperbolic, piecewise linear, and others (Zendehboudi et al., 2018). A summary of a single-neuron ANN architecture is demonstrated in Fig. 10.2. Also, Fig. 10.3 presents another conceptual model of an ANN and Ensemble architecture indicating the various stages of operation. The mathematical model of the single neuron computation with an x input is shown below

$$j = \sum_{i=1}^{n} w_i x_i + a \qquad (10.8)$$

where x_i is the input signal from the axon terminal arriving at synapse i; w_i represents the weight of the synapse i; a is the bias term, and j is the summation of all input received from the dendrites. The output signal after is given by the equation:

$$y = f(j) \qquad (10.9)$$

where y represents the neuron output signal transmitted through an axon to another perceptron, and f represents the activation transform function.

The various categories of the ANN and the activation function employed on the input variable give room for the possible design of an ANN as shown in Fig. 10.4. Zendehboudi et al. (2018) presented in detail some of the possible and popular ANN designs.

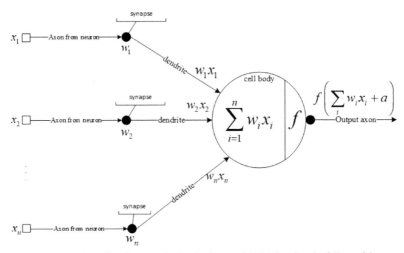

Fig. 10.2. Mathematical framework of a single neural ANN showing the follow of data.

Petroleum Reservoir Characterisation 217

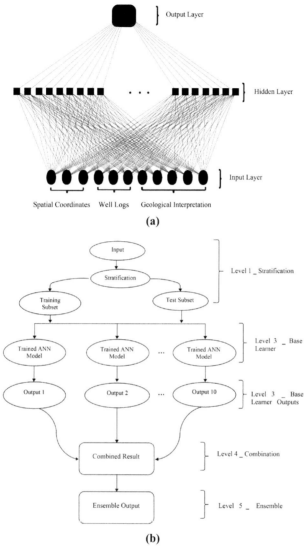

Fig. 10.3. Example of (a) conceptual model for the BPNN modelled by Mohaghegh et al. (1996), and (b) conceptual model framework for the Ensemble Learning Framework modelled after Anifowose et al. (2017).

In Fig. 10.3a, the ANN was designed for estimating petrophysical properties. The model is a three-layer BPNN with 11 inputs, 28 hidden layers, and 1 output neuron. The model utilised the sigmoid transfer functions in the hidden and output layers. The authors used a small learning rate of 0.1 and a learning momentum of 0.6 to solve the problem of nonlinearity and complexity in the input data. In Fig. 10.3b, for the ensemble work frame, level

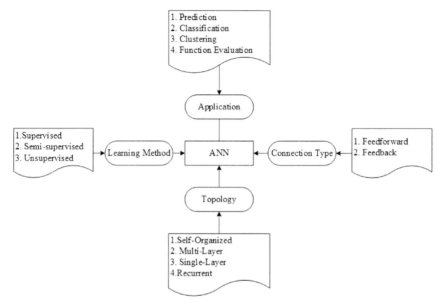

Fig. 10.4. Types of ANN based on different criteria (modified after Zendehboudi et al., 2018).

1_stratification starts by applying machine learning to identify the training and testing dataset; with level 2, a base learner is created to cater for uncertainties existing in the dataset; level 3 produces an output on the various uncertainties handled in level 2; following level 3, the performance of the trained model is accessed using the training and testing data; finally, the output is combined using suitable combination rules (level 4) to develop an overall compromise-based ensemble result (level 5).

4.1.1 ANN Application in Petrophysical Reservoir Prediction

Mohaghegh et al. (1996) introduced a new application of ANN to characterise the heterogeneity of a reservoir. They encompassed the application of a three-layer feed-forward Back Propagation Neural Network (BPNN) approach (Fig. 10.3a). The authors proposed using a General Regression Neural Network (GRNN) to first identify the optimal network design to achieve a steady state in a shorter time. The GRNN lacks reproducibility with different datasets when trained with a different dataset. Therefore, the GRNN was used by the authors to identify the optimal network design structure and the BPNN was used to obtain the final results. Upon the high reservoir heterogeneity, a good fit was achieved by comparing it with the actual laboratory measurement from core logs. Hamada and Elshafel (2010) presented a successful implementation of the ANN to determine the Phi and K of the sand gas reservoir by exploiting the transverse relaxation time of Nuclear Magnetic Resonance (NMR) T_2 and well logs. They used the ANN model which involved the NMR T_2 spin value as well as RHOB and resistivity logs to predict the Phi and K of the test wells. The ANN paradigm showed a good correlation with the core Phi and K values and that of the NMR predicted Phi and K for the test wells. Singh et al. (2016) employed the well-known BPNN to predict Phi for a gas field. Three well logs (RHOB, sonic (DT), and deep resistivity log) were used as the input variable for Phi estimation using one hidden layer. The authors correlated the predicted Phi to that of the Phi estimated using an empirical approach which entails the use of Neutron (NPHIL) and RHOB log. A correlation of 0.97 was obtained which indicates a good prediction. The robustness of the model was validated by testing against other offset wells using core measurement. The model reached a good correlation coefficient of 0.994 for the Phi estimated.

Hele and Bhatt (2002) also achieved excellent results when the authors used ANN to predict fluid saturation using wireline data while Elshafei and Hamada (2009) successfully estimated reservoir properties achieving an excellent correlation to core data using ANN. Al-Bulushi et al. (2009) applied ANN to estimate a nonlinear Sw using open-hole log data for a sandstone reservoir. From their results, they established a good Root Mean Square Error (RMSE) and correlation using the wireline logs. Zabihi et al. (2011) similarly designed an ANN model to predict K damage caused by the accumulation of barite sulfate scale in core voids with good precision during core flooding. The model took into account data from a case study to examine the K reduction owing to the build-up of barium sulfate in Malaysia and Berea sandstone cores. Results confirmed a high accuracy prediction of the modelled ANN to the measured permeability values. Nyein et al. (2018) developed an ANN paradigm for Sw prediction that included five (5) input parameters (GR, ILD, RHOB, photoelectric factor (PEF), and NPHI) and multiple hidden layer nodes. Likewise, the Mean Square Error (MSE) for the model's training and testing was 0.0109 and 0.012, respectively. This model achieved a decent correlation with core information. Likewise, the authors also utilized ANN in predicting Phi and classified the various lithology from the well logs. The model applied for estimating Phi inculcated six wireline logs (GR, ILD, RHOB, NPHI, PEF, and DT) as inputs with two latent nodes. The resulting Phi estimated matched with core data with MSEs of 0.001412 and 0.00154 for training and testing, respectively. The good prediction highlights the importance of using impactful data input to train the ANN algorithm for a specified output variable (Table 10.2).

Table 10.2. Literature on the application of ANN in petrophysical reservoir prediction.

Author(s)	Input Variables	Predicted Output	Model Used	Activation Functions	Output	Statistical Measures	Remarks
Mohaghegh et al. (1996)	Digitized well logs	Porosity, Permeability, Porosity, and Saturation	General Regression Neural Network (GRNN), Three-Layer Feed Forward Backpropagation	Sigmoid Transfer Function	Application of the feed-forward ANN proved to be successful in the prediction of the petrophysical properties	Not Noticed (NN)	For the benchmark ANN algorithm, consideration for automatic parameter selection would have helped to improve the predictive performance of the model.
Wong et al. (2005)	Well logs (GR, ILD, LLS, RXO, RHOZ, NPHI, PEF, DT)	Porosity	Support Vector Machines (SVMs), Back Propagation Neural Network (BPNN)	Gaussian Radial Basis Kernel Function	Results showed that SVM produced good results as compared to BPNN by taking into consideration the choice of the activation function	**MSE Normalised Data** SVM = 0.026347 BPNN = 0.074090 **MSE Raw Data** SVM = 23.7126 BPNN = 66.6817	Although the model achieved a good result, a more robust feature extraction or selection technique could have a better-improved prediction accuracy
Al-Bulushi et al. (2009)	Well logs (GR, NPHI, RHOZ, ILD), Cation Exchange Capacity data	Water Saturation	Artificial Neural Network (ANN) with Resilient Propagation, Multiple Linear Regression, Saturation Height Function (SHF)	Tan-Sigmoid	Results showed the robustness of saturation prediction for the test field and other fields	**ANN** RMSE = 0.025, R = 0.91 **SHF** RMSE = 0.079	Proper hyperparameter tuning is a problem and with the SHF a clear water leg should be defined in other to increase its performance.

Petroleum Reservoir Characterisation 221

Zabihi (2012)	Case Study data (Core Length, Core Diameter, Average Porosity, Permeability, Temperature, Differential Pressure)	Permeability	Multilayer Neural Network with Back Propagation Algorithm	Hyperbolic Tangent and Logistic Sigmoid	Results indicated the proposed Neural Network predicted permeability with high efficiency and a minimal error of about 2%	A Total Average Absolute Deviation (TAA) of 1.06% for testing and a TAAD of 2.03% for performance.	With a proper hyperparameter tunning algorithm, the efficiency of this algorithm will improve.
Al-Anazi and Gate (2012)	Well logs (Gr, NPhi, DT, RHOB, ILD)	Porosity and Permeability	Support Vector Regression (SVR) and Multilayer Perception (MLP) Neural Network	Gaussian radial basis and sigmoid kernel function	SVR outperformed MLP with the sparse dataset	SVM attained a 95% confidence median MSE of the prediction as compared to MLPNN	Although SVM performed well, a more robust feature selection tool, such as the embedded model, may have enhanced prediction accuracy
Gholami et al. (2012)	Well Logs (DT, GR, NPHI, RHOZ, PEF, ILD, LLS), Core data	Permeability	General Regression Neural Network (GRNN) and Support Vector Machine (SVM)	Independent Component Analysis (ICA), loss Function, kernel type k, Radial Basis Function (RBF)	After comparing results for SVM and GRNN, the SVM was more accurate than the prediction using the GRNN	SVM attained an R (0.96) and RMSE (0.28) between core and predicted permeability	The hold-out cross-validation approach used has irregular data splitting, which can lead to an incomplete model evaluation

Table 10.2 contd. ...

...Table 10.2 contd.

Author(s)	Input Variables	Predicted Output	Model Used	Activation Functions	Output	Statistical Measures	Remarks
Sebtosheikh and Salehi (2015)	Seismic Inversion, Well logs	Lithology	**Support Vector Machine (SVM)**: Statistical Learning Theory (SLT), Structural Risk Minimization (SRM), Empirical Risk Minimization (ERM)	Kernel Function	Results indicate a good prediction of the SVM after the training set size reduced	NN	With the SVM, in a case where the quantity of features for an individual data point exceeds the number of training datasets, the SVM will underperform
Anifowose et al. (2015)	Porosity and Permeability datasets were obtained from two major oil and gas reservoirs with diverse geological formations	Porosity and Permeability	Extreme Learning Machine (ELM), Functional Network, Random Forest (RF), Multivariate Regression	Sigmoidal Activation Function, Sine Activation Function and Hardlim Activation Function	Results indicate that the ELM outperformed the other machine learning algorithms with a higher correlation coefficient and lower prediction errors	The proposed ELM had average R^2 = 0.95 and 0.92; RMSE = 0.18 and 0.56; MAE = 0.08 and 0.48 for porosity and permeability, respectively.	The improved ensemble learning model adopted has been reported to be efficient in handling most classification and regression issues with greater success in the related sciences than the other algorithm used

Srisutthiyakorn (2016)	2D/3D Binary Segmented Images, Porosity, Specific Surface Area	Permeability	Multilayer Neural Network (MNN), Convolutional Neural Network (CNN), Gradient Descent (GD)	Tan-sigmoid Function, Rectified Linear Function	Results using MNN and CNN produced a good prediction of permeability especially in the case of 2D CNN in multiscale.	CNN had the lowest testing MSE of 2.4307*E05 and MNN has the lowest testing MSE of 2.3999*E05 in the case of the 2D in multiscale.	Although CNN has proven to be efficient and represent complex nonlinear functions, without a large dataset its efficiency becomes void.
Al-Mudhafar (2017a)	Well logs (NPhi, Vsh, Sw)	Lithofacies and Permeability	Probabilistic Neural Network (PNN), Generalised Boosted Regression Model (GBM), Multiple Linear Regression (MLR)	**GBM** Binomial and AdaBoost Loss Function	Results indicated a 95.81% correct prediction of lithofacies whiles GBM led to much more accurate permeability prediction compared to MLR	PNN attained a 95.81% accuracy for lithology prediction. **Permeability** *GBM* $R^2 = 0.9953$ and RMSE = 28.43 *MLR* $R^2 = 0.9551$ and RMSE = 88.42	PNN comparatively is insensitive to outliers, but with a good feature selection tool, the performance of the algorithm would have improved.
Al-Mudhafar (2017b)	Well logs (NPh, Sw, Vsh)	Lithofacies	Kernel Support Vector Machine (KSVM) and Linear Discriminate Analysis (LDA)	Non-zero Lagrangian Multipliers	KSVM had good accuracy for the lithofacies classification	KSVM lead to attaining a 99.55% accuracy for lithofacies prediction	Although LDA can recognize patterns with known prior assumptions, KSVM has proven to be efficient in identifying distinct features with the decision function defined by assisting vectors

Table 10.2 contd. ...

...Table 10.2 contd.

Author(s)	Input Variables	Predicted Output	Model Used	Activation Functions	Output	Statistical Measures	Remarks
Moghadasi et al. (2017)	Well logs (GR, NPHI, DT, Calliper, ILD, RHOZ, Sw, Vsh), lithology, Core data	Porosity and Permeability	Artificial Neural Network (ANN), Principal Component Analysis (PCA)-ANN, Statistical ANN based on bagging approach	Levenberg-Marquardt Optimization Algorithm	Results indicated that the statistical ANN had an improved quality as compared to those obtained from both PCA-based ANN and the classical ANN models.	**Classical Neural Network** *Porosity* $R^2 = 73.7\%$, MSE = 0.0089 *Permeability* $R^2 = 72.2\%$, MSE = 0.0074 **PCA based ANN** *Porosity* $R^2 = 79.8\%$, MSE = 0.0085 *Permeability* $R^2 = 74.1\%$, MSE = 0.0092 **Bagged ANN** *Porosity* $R^2 = 94\%$, MSE = 0.001 *Permeability* $R^2 = 85\%$, MSE = 0.0034	Even though the bagging ANN approach can optimize the ANN model, its performance is heavily dependent on the decision criteria, which is often difficult to determine

Petroleum Reservoir Characterisation 225

Iturraran-Viveros and Munoz-Garcia (2018)	Reflection 2D Seismic data, Seismic Attribute, Well logs (SP, ILD, DT, GR)	Porosity (Phi), Saturation (Sw), and Volume of Clay (Vsh)	Backpropagation ANN, Conjugate Gradient ANN, BFGS ANN	Gamma Test	Results show a good correlation between the predicted parameters with the core data obtained from the location	NA	With the ANN having the advantage of not knowing the process responsible for generating the data but still performing better, it always has the issue of hyperparameter tuning and also dimensionality issues.
Zhong and Carr (2019)	Core data, Well logs (GR, RHOZ, Vs1)	Porosity	Support Vector Machine (SVM), Multilayer Perceptron Neural Network (MLPNN)	Particle Swarm Optimization (PSO) and Mixed Kernel Function (MKF)	PSO-MKF-SVM displayed better performance over different methods with MLPNN having the worst.	**PSO-MKF-SVM** R = 0.9560, R^2 = 0.9140, RMSE = 1.6505, AAE = 1.4050 and MAE = 2.717	The hybridised model has highlighted a good accuracy based on prediction, reliability and computation time than the conventional algorithms
Gogoi and Chatterjee (2019)	2D Post-Stack Seismic data, Well logs (DTs, RHOZ, GR, LLS, ILD, NPHI)	The volume of shale (Vsh), Water Saturation (Sw)	Multilayer Feed Forward Neural Network (MLFN)	Sigmoid Activation Function	Results indicate a reliable prediction using the MLFN model	**Vshale** Cross-correlation of 0.88 was attained by MLFN between the actual and the predicted with a 0.031 error **Sw** A cross-correlation of 0.85 was attained by MLFN between the actual and the predicted with a 0.05 error	With the ability to approximate complex function, its hyperparameters selection becomes challenging and always rely on a trial and error approach

Table 10.2 contd. ...

...Table 10.2 contd.

Author(s)	Input Variables	Predicted Output	Model Used	Activation Functions	Output	Statistical Measures	Remarks
Soleimani et al. (2020)	Reflection Seismic data, Well Logs (RHOZ, GR, DT)	Porosity	Multi-layer Feed-Forward Network (MLFN), Radial Basis Function (RBF), Probabilistic Neural Network (PNN), Geostatistics, Regression Analysis	Gaussian Function	Results indicate the supremacy of the Neural Network compared with the other methods used	PNN had an average Validation error of 4.6 as compared to 4.8 for MLFN	With PNN having a good validation error, it is inefficient in terms of memory usage.
Okon et al. (2021)	Depth interval, GR, RHOZ, Resistivity logs	Porosity, Permeability and Water Saturation	Multiple-input and Multiple-outputs Feed-forward Back-propagation Neural Network (MIMO-FFBPNN)	Levenberg-Marquardt, Bayesian Regularisation and Scaled Conjugate Gradient	Results indicate a reliable prediction using the MIMO-FFBPNN model	**MIMO-FFBPNN Porosity** R= 0.9243 and MSE = 1.7243 **MIMO-FFBPNN Permeability** R= 0.9810 and MSE = 0.0003 **MIMO-FFBPNN Water Saturation** R= 0.9631 and MSE = 0.0049	

Mohamed et al. 2022	T_2 logarithmic mean, T_2 peak, T_2 components range and T_2 component range index	Permeability	Feed-forward Backpropagation Neural Network, Cascade-forward Neural Network, Elman Neural Network, Pattern Recognition Neural Network and Distributed Delay Neural Network	Bayesian Regularisation	Results show that the Elman Neural Network had the best prediction of permeability	**The model had the highest prediction of 0.99 correlation coefficient and 0.88 R2 as compared with the other methods**

4.2 Support Vector Machine (SVM)

The concept of support vector machines was first introduced in the early 1960s but later in the 1990s, its application became popular since it was able to outperform the classification problems of the ANN approach (Zendehboudi et al., 2018). SVMs are ML algorithms with kernel functions used for regression (SVR; Fig. 10.5a) and classification (SVC; Fig. 10.5b) analysis as shown in Fig. 10.5. SVMs have contributed widely to various applications such as regression, classification and forecasting. Some major advantage of SVMs is their ability to handle noisy data and high dimensional data (Furey et al., 2000). Also, the simplest structure of SVMs is capable of classifying two linearly-conceivable sets of data classes taking into consideration linear hyperplanes with maximum margins from both data classes.

Considering a classification SVM with 2-dimensional space class (class 1 and 2) that can be handled as a linear classification as shown in Fig. 10.5b. Assuming an n input data represented by $a = \{a_1, a_2, ..., a_n\}$ a tagged variable y_n which falls between -1 and $+1$ and can help classify the input variable. In this case, the ith input can fall into class 1 when $y_n = +1$ and class 2 when $y_n = -1$. Given a vector w which is perpendicular to the decision boundary as shown in Figure, the decision function can be expressed as:

$$c(a,w,e) = w.a + e = \sum_{i=1}^{n} w_i a_i + e \tag{10.10}$$

where c is the decision function, a represents the input vector, w denotes a vector perpendicular to the separating line, and e is a bias.

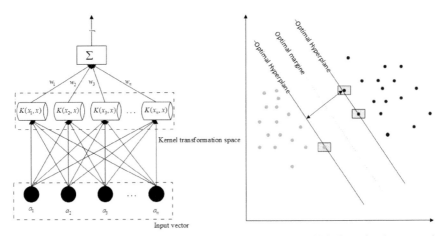

Fig. 10.5. SVM architecture showing: (a) regression SVM/RVM in a high dimensional space; and (b) supervised support vector creating a linear relationship between two non-linear data making use of optical hyperplane.

SVM classification problems correspond to the maximisation of the margins separating the two classes (Zendehboudi et al., 2018). To help minimise this, a Lagrange multiplier ($a_i \geq 0$) is introduced to convert from a constrained to an unconstrained problem. Also, SVM can classify any given input in higher dimensional space utilising a transformation such as sigmoid, polynomial, and radial basis functions in the kernel when the classes are not linearly separable using a linear hyperplane (Zendehboudi et al., 2018). Considering the kernel function, a vector x_i and x_j are introduced into the objective function. Since the optimisation depends on the dot product of the sample vector, the conversion to feature space can be abridged to find a kernel function $K(x_i, x_j)$. The mathematical expression for the objective function can be expressed as;

$$G = \sum_{i=1}^{n} \alpha_i - \frac{1}{2} \sum_{i=1}^{n} \sum_{j=1}^{n} y_i \alpha_i K(x_i, x_j) y_j \alpha_j \qquad (10.11)$$

4.2.1 Machine Learning (ML) Application in Petrophysical Reservoir Prediction

With the use of SVMs, Al-Anazi and Gate (2012) analysed the contribution of limited well log data proportion on the prediction of Phi and K in a heterogeneous sandstone reservoir. For validation, the Support Vector Regression (SVR) of the SVM was correlated with Multilayer Perception (MLP) neural network, with the output indicating the supremacy of the SVR over the MLP.

Wong et al. (2005) also predicted Phi for an offshore Western Australian heterogeneous reservoir, implementing the use of the Gaussian radial basis activation function SVM. The model does not define a relative balance between the complexity of trade-off parameter management and the observed error to minimise processing time. The authors validated their outcome by comparing the model to that of BPNN, establishing the benefit of SVM as a substitute supervised ML paradigm for characterising heterogeneous reservoirs. Gholami et al. (2012) as well did a comparative study by predicting K for gas wells in the Southern Pars field using SVM and GRNN. They employed the use of Independent Component Analysis (ICA) for the selection of dependent logs. The results showed a good correlation of 0.97 between core and predicted permeability. Also, upon comparing the results for SVM to that for GRNN, the SVM reached a faster and more accurate hydrocarbon reservoir permeability prediction than the GRNN. SVM as reported by Asante-Okyere et al. (2020b) has seen a greater performance in the field of petrophysical properties. Table 10.3 provides the applications of SVM in the field of petrophysical properties prediction.

Table 10.3. Literature on the application of SVM in petrophysical reservoir prediction.

Author(s)	Input Variables	Predicted Output	Model Used	Activation Functions	Output	Statistical Measures	Remarks
Wong et al. (2005)	Well logs (GR, ILD, LLS, RXO, RHOZ, NPHI, PEF, DT)	Porosity	Support Vector Machines (SVMs), Back Propagation Neural Network (BPNN)	Gaussian Radial Basis Kernel Function	Results showed that SVM produced good results as compared to BPNN by taking into consideration the choice of the activation function	**MSE Normalised Data** SVM = 0.026347 BPNN = 0.074090 **MSE Raw Data** SVM = 23.7126 BPNN = 66.6817	Although the model achieved a good result, a more robust feature extraction or selection technique could have a better-improved prediction accuracy
Anifowose et al. (2011)	Well Logs (GR, Phi, RHOB, Sw, ILD, CALI)	Porosity and Permeability	Functional Networks (FN), Type-2 Fuzzy Logic System (T-2 FLS), Support Vector Machine (SVM)	Kernel Activation Function	The result showed a higher correlation coefficient for the hybrid model compared with the individual models	The hybrid model attained a maximum R^2 of 0.96 for porosity prediction and R^2 of 0.70 for permeability prediction.	Although SVM can handle small datasets, the paper has proven hybridising the model has a greater potential of handling the issues of the benchmark SVM algorithm
Al-Anazi and Gate (2012)	Well logs (Gr, NPhi, DT, RHOB, ILD)	Porosity and Permeability	Support Vector Regression (SVR) and Multilayer Perception (MLP) Neural Network	Gaussian radial basis and sigmoid kernel function	SVR outperformed MLP with the sparse dataset	SVM attained a 95% confidence median MSE of the prediction as compared to MLPNN	Although SVM performed well, a more robust feature selection tool, such as the embedded model, may have enhanced prediction accuracy

Petroleum Reservoir Characterisation 231

Gholami et al. (2012)	Well Logs (DT, GR, NPHI, RHOZ, PEF, ILD, LLS), Core data	Permeability	General Regression Neural Network (GRNN) and Support Vector Machine (SVM)	Independent Component Analysis (ICA), loss Function, kernel type k, Radial Basis Function (RBF)	After comparing results for SVM and GRNN, the SVM was more accurate than the prediction using the GRNN	SVM attained an R (0.96) and RMSE (0.28) between core and predicted permeability	The hold-out cross-validation approach used has irregular data splitting, which can lead to an incomplete model evaluation
Anifowose et al. (2013)	Reflection Seismic data, Well Logs	Permeability	Support Vector Machine (SVM, Type-2 Fuzzy Logic (T2FL)	Kernel Function	Upon comparison, the SVM model gave the most accurate permeability prediction	SVM produced a good prediction for the seismic plus log input with an RMSE of about 0.49, R of about 0.7, and MAE of about 0.38	SVM outperforms T2FL when working with small data size as compared to T2FL which work best with input data with uncertainties
Sebtosheikh and Salehi (2015)	Seismic Inversion, Well logs	Lithology	**Support Vector Machine (SVM)**: Statistical Learning Theory (SLT), Structural Risk Minimization (SRM), Empirical Risk Minimization (ERM)	Kernel Function	Results indicate a good prediction of the SVM after the training set size reduced	NN	With the SVM, in a case where the quantity of features for an individual data point exceeds the number of training datasets, the SVM will underperform

Table 10.3 contd. ...

...Table 10.3 contd.

Author(s)	Input Variables	Predicted Output	Model Used	Activation Functions	Output	Statistical Measures	Remarks
Al-Mudhafar (2017b)	Well logs (NPhi, Sw, Vsh)	Lithofacies	Kernel Support Vector Machine (KSVM) and Linear Discriminate Analysis (LDA)	Non-zero Lagrangian Multipliers	KSVM had good accuracy for the lithofacies classification	KSVM lead to attaining a 99.55% accuracy for lithofacies prediction	Although LDA can recognize patterns with known prior assumptions, KSVM has proven to be efficient in identifying distinct features with the decision function defined by assisting vectors
Zhong and Carr (2019)	Core data, Well logs (GR, RHOZ, Vsh)	Porosity	Support Vector Machine (SVM), Multilayer Perceptron Neural Network (MLPNN)	Particle Swarm Optimization (PSO) and Mixed Kernel Function (MKF)	PSO-MKF-SVM displayed better performance over different methods with MLPNN having the worst.	**PSO-MKF-SVM** $R = 0.9560$, $R^2 = 0.9140$, RMSE = 1.6505, AAE = 1.4050 and MAE = 2.717	The hybridised model has highlighted a good accuracy based on prediction, reliability and computation time than the conventional algorithms
Wang et al (2019)	Core data, Wireline logs (GR, Vsh)	Microfacies	Support Vector Machine (SVM)	Radial Basis Kernel Function	SVM algorithm attained a good automatic and quantitative identification of depositional microfacies with a good prediction accuracy	Results indicate an 84% prediction accuracy attained using the SVM model	Although the paradigm had a high prediction accuracy, it could have been higher if a good feature selection tool had been used to identify the relevant parameters for model training.

Qiang et al. (2020)	Well Logs (GR, SP, Calliper, ILD, DT, DTs, RHOZ, NPHI), Core Data, and Post stack Inversion	Porosity, Saturation, Spatial distribution of acoustic impedance	Support Vector Machine (SVM), Sequential Gaussian Simulation (SGS), and Gaussian Indicator Simulation (GIS)	Radial Basis Kernel Function	SVM Showed good prediction for porosity while a hybrid of GIS and SVM gave a higher prediction of spatial variation of reservoir properties for the reservoir	SVM attained an average overall R of 0.84 for all the input wells	The implementation of feature selection tools may have resulted in the improvement of the performance of the model.
Xie et al. (2020)	Well Log (GR, CNL)	Lithology Identification	Randomised Ensemble Tree Classifier (RETC), Random Forest (RF), Gradient Tree Boosting (GTB), and Xgboosting	N-fold Cross-Validation, Multivariate Outlier	Results show that the proposed framework (RETC) has a higher prediction accuracy for sandstone lithology than the other AI classifiers.	RETC had a prediction accuracy of 89.4% and 91.1% in the Daniudui and Hangyinqi Gas Field, respectively	With the data input considered for the lithology identification, the prediction accuracy would have improved if another contributing well log such as photoelectric factor, and density were used.
Asante-Okyere et al. (2020a)	Core data, Well logs (GR, DT, SP, ILD, NPHI)	Water Saturation	Least Square Support Vector Machine (LSSVM), Subtractive Clustering Method (SCM), Adaptive Neuro-fuzzy inference System (ANFIS)	Principal Component Analysis (PCA), Coupled Simulation Annealing (CSA), Simplex Algorithm	Overall PCA-LSSVM had the least prediction error and outperformed PCA-ANFIS-SCM for the prediction of saturation.	**PC-LSSVM** Testing (RMSE and AAE) = 0.048 and 0.034 **ANFIS-SCM** Testing (RMSE and AAE) = 0.055 and 0.036	The hybridised SVM attained a good prediction due to the application of the feature selection technique adopted and its ability to also work perfectly with a small dataset

Petroleum Reservoir Characterisation 233

5. Lithology and Facies Analysis

The study of sedimentary depositional environment and its heterogeneity (Table 10.4) has been studied by many authors (e.g., Meshri, 1986; Pettijohn, 1983; Yu, 2008) except Qui (2000). The various authors suggested that the sedimentary depositional environment and its heterogeneity comparison can be best classified by studying its facies. In the early 1990s, Walker and James (1992) underlined that facies analysis is essential when studying basin fills. Lithology and facies analysis help to subdivide the reservoir rock sequence into respective lithostratigraphic units to help map and correlate between depositional areas and analyse their source. Due to the complexity of facies modelling, Tang (2008) outlined the two basic approaches for facies modelling, which are: the exploitation of empirical equations such as the Archie equation, multivariate statistics, and neural network. In recent years, the latter has gained more attention, with which several researchers have used multivariate statistics and neural networks for facies modelling (e.g., Bestagini et al., 2017; Miller et al., 2019; Tran, 2020) and identify lithology (e.g., Al-Anazi and Gates, 2010; Al-Mudhafar, 2017a; Alzubaidi et al., 2021).

5.1 AI Applications in Lithology and Facies Analysis

Rafiq et al. (2016) proposed an idea of microseismic facies analysis that enables the segmentation of an eccentric reservoir into discrete facies in response to microseismic accompanied by the combined exposition of microseismic with 3D seismic data. This approach was originally proposed by Eaton et al. (2014) and has been modified by the authors. Their findings correlated with the principal curving irregularities from 3D seismic data. With the combination of microseismic facies analysis and field parting acquired from log and core data, the authors delineated reservoir segments marking structural and depositional trends. De Matos et al. (2007) proposed two new semiautomatic alternative methods for seismic facies analysis. They used Kohonen Self-Organising Maps (SOMs) clustering as an advanced approach for mapping seismic facies, evaluating their quantity, and selecting seismic trace peculiarities in each oriented geological segment with the help of the wavelet transform algorithm. Validating their results, both approaches were used and utilised the synthetic and real seismic data from the Namorado deepwater oilfield in Campos Basin, offshore Brazil. For several seismic data sets, the results confirmed a good prediction of SOM clustering. The results also showed an improved seismic facies analysis by utilising trace distinctiveness noticed by the wavelet transform technique.

An integration of wireline logs into seismic lithofacies classification was conducted by Avseth and Mukerji (2002). The input data for the study comprised GR, RHOB, seismic velocity, and impedance. The lithofacies

Table 10.4. Literature on the application of AI in lithology and facies analysis.

Author(s)	Input Variables	Predicted Output	Model Used	Activation Functions	Output	Statistical Measures	Remarks
Avseth and Mukerji (2002)	Well logs (GR, RHOZ, DT)	Lithofacies	Rock Physics, Mahalanobis Discriminant Analysis (MLDA), Probability Density Function (PDF), Classical Neural Network (NN)	MLDA: Mean and Covariance. NN: Sigmoid Transfer Function	Results demonstrate a somewhat better forecast utilizing the NN as compared with the PDF and the MLDA, yet the MLDA ended up being viable for the grouping of well log into discrete lithofacies	All applied models had a success rate of about 80%	The predictive performance of the model would have been improved if additional inputs such as photoelectric factor and spontaneous potential were considered
Tang (2008)	Well Logs (GR, SP, DT, NPHI, RHOZ, PE, ILD)	Electrofacies	Probabilistic Neural Network (PNN)	Radial Basis function	PNN had a good classification performance	PNN attained a prediction accuracy above 70%	Although the PNN achieved a good classification, comparing it with other statistical classification tools would have indicated the prediction strength of the algorithm as compared to other models.
Al-Mudhafar (2017a)	Well logs (NPhi, Vsh, Sw)	Lithofacies and Permeability	Probabilistic Neural Network (PNN), Generalised Boosted Regression Model (GBM), Multiple Linear Regression (MLR)	GBM Binomial and AdaBoost loss Function	Results indicated a 95.81% correct prediction of lithofacies whiles GBM led to much more accurate permeability prediction compared to MLR	PNN attained a 95.81% accuracy for lithology prediction. Permeability GBM $R^2 = 0.9953$ and RMSE = 28.43 MLR $R^2 = 0.9551$ and RMSE = 88.42	PNN comparatively is insensitive to outliers, but with a good feature selection tool, the performance of the algorithm would have improved.

Table 10.4 contd....

...Table 10.4 contd.

Author(s)	Input Variables	Predicted Output	Model Used	Activation Functions	Output	Statistical Measures	Remarks
Al-Mudhafar (2017b)	Well logs (NPhi, Sw, Vsh)	Lithofacies	Kernel Support Vector Machine (KSVM) and Linear Discriminate Analysis (LDA)	Non-zero Lagrangian Multipliers	KSVM had good accuracy for the lithofacies classification	KSVM lead to attaining a 99.55% accuracy for lithofacies prediction	Although LDA can recognize patterns with known prior assumptions, KSVM has proven to be efficient in identifying distinct features with the decision function defined by assisting vectors
Wang et al (2019)	Core data, Wireline logs (GR, Vsh)	Microfacies	Support Vector Machine (SVM)	Radial Basis Kernel Function	SVM algorithm attained a good automatic and quantitative identification of depositional microfacies with a good prediction accuracy	Results indicate an 84% prediction accuracy attained using the SVM model	Although the paradigm had a high prediction accuracy, it could have been higher if a good feature selection tool had been used to identify the relevant parameters for model training.
Ameur-Zaimeche et al. (2020)	Well Logs (GR, RHOZ, K, TH), Core data	Lithofacies	Multilayer Perceptron Neural Network (MLPNN), Classical Cluster Analysis (CCA)	MLPNN Sigmoid Activation Function CCA Partitioning and Hierarchical	MLPNN had a good lithofacies prediction with the lowest RMS and highest R-value as compared with the classical cluster analysis	MLPNN RMSE and MAE = 0.39 and 0.23, respectively R (testing) = 0.92 CCA RMSE and MAE = 1.04 and 0.54, respectively R = 0.68	MLPNN is insensitive to outliers present in the input data and also work well for complex function as compared with CCA which is sensitive to outliers.

Xie et al. (2020)	Well Log (GR, CNL)	Lithology Identification	Randomised Ensemble Tree Classifier (RETC), Random Forest (RF), Gradient Tree Boosting (GTB), and Xgboosting	N-fold Cross-Validation, Multivariate Outlier	Results show that the proposed framework (RETC) has a higher prediction accuracy for sandstone lithology than the other AI classifiers.	RETC had a prediction accuracy of 89.4% and 91.1% in the Daniudui and Hangyinqi Gas Field, respectively	With the data input considered for the lithology identification, the prediction accuracy would have improved if another contributing well log such as photoelectric factor, and density was used.
Asante-Okyere et al. (2020b)	Lithology data and Well logs (RHOB, GR, SP, CN, and ILD)	Lithofacies	Gradient-Boosted Machine (GBM)	K-mean, Gaussian Mixture Models	Results show improved performance in terms of classification accuracy by the K-mean-based GBM	Comparing the various model GBM had a Training and Testing accuracy of 80.97% and 80.96%, respectively while K-mean GBM had 82.89% and 81.42% for training and testing, respectively	The K-mean cluster algorithm employed is based on the cluster centres of the data with which if the data cluster does not show a defined circular cluster, its classification becomes difficult.
Chai et al. (2022)	Seismic data	Seismic facies	Deep Learning (DL) Neural Network	BridgeNet and ResNet	The accuracy of projected facies along the training and testing portions indicates that the results are congruent with geologic sedimentation, confirming the generalization capabilities of the enhanced flexible BridgeNet. DL	NN	An ANN with a higher kernel size outperforms one with a shorter kernel size. Furthermore, the BridgeNet with shortcut connections outperforms the BridgeNet without shortcut connections.

classification was based on the six different facies group from the North Sea turbidite reservoir, taking into consideration the clay content, grain size, and bedding configuration. After data filtering, the authors used a neural network and various multivariate statistical techniques to classify the lithofacies. The Mahalanobis Linear Discriminant Analysis (MLDA), Probability Density Function (PDF) classifier, and a Multilayer Feed-forward design with Back-propagation (MLFN-BP) mass adaptation methodology were used based on their ability to establish a link between facies and their physical properties. The results emphasised the use of PDF for the identification of new facies using holdout cross-validation; however, the MLFN-BP classifier was the best algorithm for dealing with the multidimensional cluster boundary. In terms of model performance, Otchere et al. (2021a), suggested there would have been an improvement if a stronger selection technique was employed to select only pertinent input variables and outlined a selection bias for their model. To also increase the performance of their model, the rock physics model would have been the best conditioning tool for the input well log data. Ameur-Zaimeche et al. (2020) conducted additional research on the feed-forward neural network using the MLP to reconstruct the lithofacies breaks in the Sif Fatima oil field in Algeria. Linking MLP to cluster analysis, their outcomes showed that MLP is better suited for forecasting the non-cored lithofacies.

Tang (2008) defined carbonate lithofacies with some well logs using a Probabilistic Neural Network (PNN) by analysing the multidimensional correlation between the variables. Because of its ability to analyse the multidimensionality that exists between well logs and distinct facies, the PNN outperformed discriminate analysis and multi-logistic statistical algorithms. To assess the reliability of the PNN lithofacies prediction, an integration of two log lithology indicators and the model was used to map zonally using simple kriging. There was a good correlation between the zonal facies map with the conceptual reservoir model which indicates a good reservoir modelling and flow simulator. Al-Mudhafar (2017b) used well logs to investigate the use of the Kernel Support Vector Machine (KSVM) to model the distribution of lithofacies. Because of its ability to recognize distinct classes with a decision function defined by additional subgroups of supporting vectors, the author chose KSVM. To further validate the model, it was cross-validated using the known lithofacies. The validation yielded a 99.55% accuracy using the KSVM algorithm. Also in recent years, there has been increasing documentation of improved facies and lithology analysis (e.g., Asante-Okyere et al., 2020b; Otchere et al., 2022b; Shen et al., 2019; Xie et al., 2021).

6. Seismic Guided Petrophysical Property Prediction

Integration of seismic and well log data for petrophysical property prediction requires bringing the source data to a common dimension (depth and time), specifically with both source in depth (well log) and time (seismic). This is usually achieved by using a seismic-to-well tie (Saikia et al., 2020) by matching the seismic response at the well location to that of the synthetic trace extracted from the well log. The synthetic trace is generated by convolving the reflection coefficient with a synthetic wavelet and using the reflection behaviour of the geological layers. Upon generating the synthetic wavelet, information on the depth of the synthetic trace and its equivalent time for the real seismic trace would be established. Once the best connection (stretch and squeeze) is attained, the depth-to-time relation is established (Saikia et al., 2020).

The use of seismic-guided reservoir property prediction is one of the improved methods to help estimate reservoir properties (Table 10.5). In fields where there are not many wells and interpreting the reservoir only from the well logs becomes difficult, this method adopts statistical tools to estimate the reservoir property for the entire field. The information gained from the well log is not used proficiently for a larger field since this information will be confined to a specific location. Due to this difference in well log information, integrating the seismic data helps to give an estimation of the varying reservoir property of the field at large. In recent years, different AI algorithms have been implemented to help in the estimation of reservoir properties with the use of seismic data and well logs.

Gogoi and Chatterjee (2019) predicted reservoir properties from well logs and seismic data by employing the MLF algorithm in Tipam sandstone and Barail arenaceous sandstone reservoirs. The data set used includes complete acoustic impedance from 2D post-stacked seismic data, as well as RHOB, Phi, and shear impedance. The MLF algorithm generated a single-layer shale volume and Sw from numerous hidden layers. Upon analysis, the outcome showed a minimal estimation error for both reservoirs, suggesting that MFL can be used to predict such properties. The MLF has been established to be effective in the estimation of sand or shale reservoir properties by Iturraran-Viveros and Munoz-Garcia (2018). They investigated the use of seismic traits and wireline logs as input factors for describing petrophysical properties for the Tenerife field at a seismic scale, with the option of sands or shales. The gamma test, a multidimensional smoothing tool, was used to help optimise the input variable selection for training the ANN. It thus supported a decent assessment of Phi and Sw and brought about a decent connection with core, well log just as seismic amplitudes for each facies. Soleimani et al. (2020) assessed the performance of the MLF by comparing it with PNN, RBF, Multiple Linear Regression (MLR), and Geostatistical Method (GM)

Table 10.5. Literature on the application of AI in seismic guided petrophysical property prediction.

Author(s)	Input Variables	Predicted Output	Model Used	Activation Functions	Output	Statistical Measures	Remarks
De Matos et al. (2007)	3D Seismic data	Seismic Facies	Self-Organizing Maps (SOMs), Wavelet Transform, K-mean Partitive Clustering	Gaussian-distribution function	AN improvement in seismic facies prediction was obtained by using the wavelet transform technique to detect the trace singularities	NN	Even though SOM provides good lithofacies classification with seismic data as input, the addition of well log data would help improve the classification performance.
Anifowose et al. (2013)	Reflection Seismic data, Well Logs	Permeability	Support Vector Machine (SVM, Type-2 Fuzzy Logic (T2FL)	Kernel Function	Upon comparison, the SVM model gave the most accurate permeability prediction	SVM produced a good prediction for the seismic plus log input with an RMSE of about 0.49, R of about 0.7, and MAE of about 0.38	SVM outperforms T2FL when working with small data size as compared to T2FL which work best with input data with uncertainties
Sebtosheikh and Salehi (2015)	Seismic Inversion, Well logs	Lithology	**Support Vector Machine (SVM)**: Statistical Learning Theory (SLT), Structural Risk Minimization (SRM), Empirical Risk Minimization (ERM)	Kernel Function	Results indicate a good prediction of the SVM after the training set size reduced	NN	With the SVM, in a case where the quantity of features for an individual data point exceeds the number of training datasets, the SVM will underperform

Petroleum Reservoir Characterisation 241

Rafiq et al. (2016)	Post stack 3D Seismic data, seismic attributes, Well logs, Core data Hydraulic fracture treatment log	Microseismic Facies	Cluster Analysis	Principal Curvature Anomaly	Results indicate integrating microseismic with seismic data provides valuable insight for characterizing unconventional reservoirs	NN	Although cluster analysis provided great insight, it is heavily affected by sampling error and would appear to be legitimate upon analysis.
Srisutthiyakorn (2016)	2D/3D Binary Segmented Images, Porosity-Specific Surface Area	Permeability	Multilayer Neural Network (MNN), Convolutional Neural Network (CNN), Gradient Descent (GD)	Tan-sigmoid Function, Rectified Linear Function	Results using MNN and CNN produced a good prediction of permeability especially in the case of 2D CNN in multiscale.	CNN had the lowest testing MSE of 2.4307*E05 and MNN has the lowest testing MSE of 2.3999*E05 in the case of the 2D in multiscale.	Although CNN has proven to be efficient and represent complex nonlinear functions, without a large dataset its efficiency becomes void.
Iturraran-Viveros and Munoz-Garcia (2018)	Reflection 2D Seismic data, Seismic Attribute, Well logs (SP, ILD, DT, GR)	Porosity (Phi), Saturation (Sw), and Volume of Clay (Vsh)	Backpropagation ANN, Conjugate Gradient ANN, BFGS ANN	Gamma Test	Results show a good correlation between the predicted parameters with the core data obtained from the location	NN	With the ANN having the advantage of not knowing the process responsible for generating the data but still performing better, it always has the issue of hyperparameter tuning and also dimensionality issues.

Table 10.5 contd. ...

...Table 10.5 contd.

Author(s)	Input Variables	Predicted Output	Model Used	Activation Functions	Output	Statistical Measures	Remarks
Gogoi and Chatterjee (2019)	2D Post-Stack Seismic data, Well logs (DTs, RHOZ, GR, LLS, ILD, NPHI)	The volume of shale (Vsh), Water Saturation (Sw)	Multilayer Feed Forward Neural Network (MLFN)	Sigmoid Activation Function	Results indicate a reliable prediction using the MLFN model	**Vshale** A cross-correlation of 0.88 was attained by MLFN between the actual and the predicted with a 0.031 error **Sw** A cross-correlation of 0.85 was attained by MLFN between the actual and the predicted with a 0.05 error	With the ability to approximate complex function, its hyperparameters selection becomes challenging and always rely on a trial and error approach
Soleimani et al. (2020)	Reflection Seismic data, Well Logs (RHOZ, GR, DT)	Porosity	Multi-layer Feed-Forward Network (MLFN), Radial Basis Function (RBF), Probabilistic Neural Network (PNN), Geostatistics, Regression Analysis	Gaussian Function	Results indicate the supremacy of the Neural Network compared with the other methods used	PNN had an average Validation error of 4.6 as compared to 4.8 for MLFN	With PNN having a good validation error, it is inefficient in terms of memory usage.

| Qiang et al. (2020) | Well Logs (GR, SP, Calliper, ILD, DT, DTs, RHOZ, NPHI), Core Data, and Post stack Inversion | Porosity, Saturation, Spatial distribution of acoustic impedance | Support Vector Machine (SVM), Sequential Gaussian Simulation (SGS), and Gaussian Indicator Simulation (GIS) | Radial Basis Kernel Function | SVM Showed good prediction for porosity while a hybrid of GIS and SVM gave a higher prediction of spatial variation of reservoir properties for the reservoir | SVM attained an average overall R of 0.84 for all the input wells | The implementation of feature selection tools may have resulted in the improvement of the performance of the model. |

by estimating Phi and compressional wave velocity from seismic attribute and well logs in an Iranian oil field. The PNN-Gaussian function selection documented the most reduced error, most noteworthy correlation, and least computational time as compared to the others by using three unique datasets for testing, training, and validation.

Evaluation of SVM to ANN and Type-2 Fuzzy Logic (TFL) was conducted by Anifowose et al. (2013) to help improve the forecast of K in a carbonate reservoir. The prediction was done by the use of a 3D seismic signal and wireline logs as input parameters. The results showed that combining seismic signatures and wireline logs enhanced the training performance of the SVM and ANN models. Sebtosheikh and Salehi (2015) led a test impact of a limited training dataset in estimating the lithology from transformed seismic characteristics and directed log information in a heterogeneous carbonate reservoir using the SVM technique, the outcomes of which demonstrated its robustness in the face of limited data. It was also discovered that reducing the training and testing data sizes did not affect the presentation of SVM in lithological prediction. To improve the prediction of the spatial distribution of acoustic impedance, Phi, and saturation for a gas field, Qiang et al. (2020) proposed a new model to help improve the reservoir quality prediction from the integration of well logs, core and seismic inversion data. The authors employed the use of Sequential Gaussian Simulation (SGS) and Gaussian Indicator Simulation (GIS), which are both statistical algorithms to estimate the spatial distribution of the reservoir properties. In order to improve the robustness of the model, it was calibrated to post-stack seismic inversion to establish a relationship utilising SVM, Radial Basic Function (RBF) kernel, and well logs. Finally, SVM produced a good relationship between the thin lithofacies layers and seismic waveforms, resulting in a better prediction of the facies' spatial distinction. Also, there was a good correlation of Phi and Sw prediction utilising SVM to facies classification with superior reservoir zones. To show the robustness of SVM, Wang et al. (2019) applied SVM to identify depositional microfacies from wireline logs. The SVM model was trained using integrated depositional microfacies obtained via core sample observation and well logs. The authors developed a quantifiable discrimination SVM model and compare it to standard results, which aided in the prediction of depositional microfacies in uncored wells with up to 84% accuracy.

7. Hybrid Models of AI for Petrophysical Property Prediction

AI has proven to improve the accuracy of reservoir property prediction as compared to conventional mathematical approaches, yet the need to further develop its effectiveness is still sought after (Otchere et al., 2021a). The various improvements to AI model accuracy and reducing processing time are to incorporate different models to handle the various limitations such as issues of dimensionality, overfitting, and local minima. In order to curb this limitation, various literature has applied various models for feature selection (Asante-Okyere et al., 2020; Dorrington et al., 2004; Lim et al., 2004; Zerrouki et al., 2014) and improved its optimisation (Amiri et al., 2015; Anifowose et al., 2011; Anifowose et al., 2017; Saemi et al., 2007) to improve the modelling ability of AI algorithms.

Moghadasi et al. (2017) investigated the use of bagged ANN, ANN-based Principal Component Analysis (PCA), and BPNN to predict permeability and porosity. Seven well logs were utilised as model parameters from the 11 wells used, and the predicted output was trained using core permeability and porosity. PCA was used as a feature extraction method to help identify the relevant variables. In order to average the predicted multiple datasets to enhance the level of stability and lower uncertainty, the bootstrap sampling bagging technique was used to generate multiple training sets for the prediction. Based on the models generated, the bagged ANN ensemble model prediction documented the highest R^2 of 0.94 (94%) and 0.85 (85%), respectively, and the lowest MSE of 0.001 and 0.0034 for Phi and K. Srisutthiyakorn (2016) developed a Multilayer Neural Network (MNN) and Convolutional Neural Network (CNN) paradigms in predicting permeability from geometry and feature extraction from rock images. Their paradigm employed slope descent-based back-propagation with Bayesian regularisation for training. MNN was modelled using porosity, Euler number, integral of the mean curvature, and specific surface area. From their research, the feed-forward network gave rise to a higher testing MSE (mean squared error) compared to the Bayesian regularisation network and CNN.

Xie et al. (2020) developed a coarse-to-fine basis that incorporates outlier detection, and multi-class classification with a very randomized tree-based classifier to resolve the issue of repeatability of most machine learning algorithms. Comparisons with some ML classifiers, such as random forest, gradient tree boosting, and xgboosting, were used to help improve the model's competency. In sandstones, the model outperformed the baseline classifiers in prediction accuracy. Asante-Okyere et al. (2020a) adopted a hybridised model to precisely estimate reservoir water saturation. This model adopted PCA as a feature extraction technique to boost the effectiveness of the optimised least square support vector machine (LSSVM) and adaptive neuro-fuzzy inference system-subtractive cluster method (ANFIS-SCM). During training and testing the PC-LSSVM model achieved a better performance as compared to the proposed PC-ANFIS-SCM model. For their outcome, the PC-LSSVM hybridised model had the lowest prediction error and outperformed the hybridised PC_ANFIS-SCM model with the valuation of water saturation. Asante-Okyere et al. (2020b) proposed a hybrid model for lithology classification that combines gradient-boosted machine and cluster algorithms. The authors take advantage of the benefit of combining results from clustering well log data into two and three groups using K-means and Gaussian mixture models (GMM) to create a more precise gradient-boosted machine (GBM) lithology model. Their findings demonstrated improvement in terms of classification accuracy rate by the K-means-based GBM classifier. Furthermore, when the setup classifiers were tested on the entire dataset, the GMM-based GBM perceived an improved presentation. The classifiers' evaluation of the matrices revealed that the cluster-based hybrid GBM models performed well, which was attributed to an increase in quality in perceiving mudstone and siltstone, which are the primary lithofacies present in the South Yellow Ocean's Southern Basin.

Anifowose et al. (2015) investigated the performance of SVM to determine the optimal value of the regularization parameter on which it is dependent. Following further evaluation, a stacked generalization ensemble model of SVM was suggested to improve reservoir parameter forecasting while accounting for K and Phi. This study used input variables from six North American well logs and eight Middle Western well logs for K estimation. Following a comparison with the standard SVM, the suggested SVM ensemble outscored the orthodox one. To satisfy the increased demand for the hybrid model to solve complex industrial problems, Anifowose et al. (2011) hybridised the functional network and SVM (FN-SVM) to further improve the reliability of the orthodox SVM. Six wells with variable conventional well logs were used as inputs for the model to predict K and Phi and were contrasted with the conventional SVM. With the use of the Least Square Based feature selection algorithm of FN, the independent well logs were chosen, which helped with the dimensionality reduction problem and improved the time productivity of the model. By comparing the models, the FN-SVM had a higher correlation coefficient and a shorter handling time than the conventional one. Zhong and Carr (2019) compared a hybrid Particle Swarm Optimisation (PSO) mixed kernel function-based SVM (PSO-MKF-SVM) to various types of MKF-SVM, SVM, and ANN in the prediction of Phi when well logs and core data were insufficient. Gamma Ray (GR), density logs, and three extracted parameters (slope of GR, slope of density, and Volume of Shale) from six wells were used as input for the model. SVMs with different core functions have been used to predict Phi and experienced difficulties in discovering the regularisation parameter and the global optimum. Notwithstanding, the applied MKF-SVM showed an improvement in the localisation of the optimal control parameters, resulting in an increased efficiency based on R^2, correlation coefficient (r), and RMSE. To test their equivalence, the MKF-SVM (RBF-LS-SVM) was compared to MLPNN and the newly developed PSO-MKF-SVM. As a basis for comparison, the proposed PSO-MKF-SVM model outperformed the other methods with RMSE = 1.6505, R^2 = 0.9140, r = 0.9560, Average Absolute Error (AAE = 1.4050) and Maximum Absolute Error (MAE = 2.717).

Otchere et al. (2021b) proposed a hybrid ensemble model for the prediction of reservoir permeability and water saturation using well logs from the Volve Field. The input logs used were GR, calliper (CALI), RHOZ, DT, NPHI, PEF, rate of penetration (ROP) and resistivity logs from two wells. The authors applied a pre-processing tool to clean the data and select input variables which are relevant for training and prediction. The grid search hyperparameter algorithm was used to select optimal parameters for effective model training. The model results indicated a high training accuracy for the traditional XGBoost for permeability while the proposed ensemble model (RFLasso-XGBoost regression) had the most robust fit for the permeability prediction of 0.98 test score. Also, for water saturation prediction, the proposed RFLasso-XGBoost model exhibited the highest correlation coefficient for both training (0.98) and testing (0.93). Otchere et al. (2022a) proposed a new application of deep ensemble super learner to establish a relationship between wireline logs and NMR T_2LM and predict NMR T_2LM. The predicted log served as the basis for subsequent input for their developed methodology and mathematical models to predict reservoir wettability. The core plugs of sandstone samples from Western Australia were used for the analysis. Results indicate an excellent performance of the deep ensemble super learner.

8. Summary

Tables 10.2 to 10.5 show a summary of the works done concerning the application of AI techniques in reservoir characterisation. The summary comprises the input data, predicted output, AI architecture, activation function, output performance, and the statistical measure for the AI algorithm. An overview of the advantages and limitations of some AI algorithms is discussed in detail by Saikia et al. (2020). It can be observed from Tables 10.2 to 10.5 that all works done considered supervised and unsupervised AI algorithms to achieve a good prediction for their developed models. Considering this, most industries produce enough data for their operations but to confirm the actual ground truth, a small amount of core data are being generated to validate the model of the empirical relation used. With little core information, training a supervised model becomes difficult. To put to use the few core dataset, a semi-supervised algorithm becomes useful. The semi-supervised algorithm considers using the small dataset in a supervised manner and later applies an unsupervised model to optimise the final prediction. From Tables 10.2 to 10.5, it can be seen that there is limited work done concerning unsupervised techniques and hybrid unsupervised techniques in the field of reservoir characterisation. Furthermore, it has also been observed that with the various

AI algorithms used, conditioning of the input data helps to improve the prediction performance of the algorithm. The use of rock physics models as a complementary tool in conditioning the well logs has in recent times gained much attention and has proven to be useful in well log conditioning. Also, much work in combining rock physics and AI algorithms has not been done. In all, it can be inferred from Table 10.1 that, this review work will be a basis to advance the study of AI applications in reservoir characterisation.

9. Challenges and Perspectives

Yu et al. (2011) highlighted that reservoir characterisation is a multidisciplinary integration exercise that involves qualitative and quantitative interpretation. With increased complexity and abundance of data from various geological and reservoir problems, new technologies and methods are increasingly adopted in reservoir characterisation. Proper integration substitutes the need to run expensive and unrelated logs to make available correct estimation faster and strengthen your confidence as an operator. A good representative reservoir characterisation presents a good model for the optimisation of the lifetime performance of a field. Achieving that requires an accurate measurement available from multiple disciplines to prepare a good quantitative and qualitative representation of the reservoir. Significant experiments have been made in this regard, and many challenges remain, some perspectives for the future are discussed.

9.1 AI Perspective

Literature has highlighted various modifications of the ANN algorithm like the functional network (FNN), PNN, and Radial Basis Function (RBF) to help remove the various limitation of the ANN algorithms. This has led to addressing nonlinearity between various inputs which the empirical correlations could not highlight. AI has produced improved prediction and classification in a different task in reservoir characterisation, that is, the use of hybrid modelling (Amiri et al., 2015; Anifowose et al., 2011; Anifowose et al., 2017; Saemi et al., 2007) and also ensemble model (Anifowose et al., 2015) to make available good hyperparameters for modelling.

Feature extraction and selection by AI algorithms is one of the main areas most literature is addressing to help improve the prediction capability of these AI tools. Feature selection looks at the individual input for the AI algorithms by looking at the necessary and unnecessary features which intern affect the efficiency of the algorithm. Since the conventional AI algorithm lacks the power of feature extraction, most research has looked at hybridising AI algorithms to cater to that deficiency. Considering the various feature selection tools (PCA, forward feature selection (FFS), linear discriminant analysis (LDS), etc.), fuzzy-aided hybrid models have proven to provide good performance in many reviewed literature.

Due to the spatial dimension of the various dataset used to train the AI algorithm, the need to transform the dataset from a high dimensional space to a low dimensional space while keeping the high dimensional structure helps improve the prediction power of the ANN and ML algorithm. This process of reducing dimensionality is termed feature extraction. With the development of deep learning algorithms, the problem of feature extraction has been addressed since the algorithm can extract valuable information from the primary data input. The literature reviewed here has indicated that the application of soft computing methods has to do with the trial-and-error method of choosing the parameter settings to obtain high-performance accuracy. Therefore, future studies have to employ an automatic selection of parameters for soft computing methods to avoid long time wasted in looking for optimal parameters.

In recent years, the oil industry has seen a high volume of data generation by progressive sensors (Saikia et al., 2020) which has led to the development of advanced modelling and estimation techniques. Such advanced modelling is the application of a deep learning algorithm (e.g., CNN) to handle feature selection (Shaheen, 2016) and extract useful messages from the input data through its hidden neurons (Saikia et al., 2020). The application of deep learning (DL) has not been well exploited in the domain of reservoir characterisation but has yielded much success in the field of image, and speech recognition (Bae et al., 2016; Fujiyoshi et al., 2019; He et al., 2016; Pouyanfar et al., 2018) and other fields like the medical sector (Chen et al., 2018; Fang et al., 2019). Although DL has in recent years been applied to solve various learning assignments, training the model is difficult (Rere et al., 2016) but with improved data availability, the performance of DL models has been better. Various algorithms such as Stochastic Gradient Descent (SGD), Conjugate Gradient (CG), Hessian-free optimization (HFO), and Krylov Subspace Descent (KSD) have been implemented to curtail this deficiency over the years. These algorithms have shown some limitations such as several manual tuning schemes for SGD, slowness of the CG, and more memory consumption of the HFO and KSD (Rere et al., 2016). To resolve the issues of some of these algorithms, the hybridisation of metaheuristics to DL will be a good area to address. The metaheuristic optimisation techniques have been successfully applied to solve many optimization problems in the research area of engineering, sciences, and related industries. Research on hybridising DL with metaheuristics algorithms is scarce in the field of reservoir characterisation. Notwithstanding, its other important status which is learning from unlabelled data concerning reservoir characterisation has also not been properly exploited. Likewise, with the availability of core information coupled with well log data, the application of semi-supervised deep learning approaches can lead to greater success in

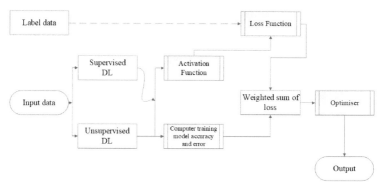

Fig. 10.6. Proposed framework for the semi-supervised learning Deep Neural Network.

reservoir characterisation and has to be exploited. Therefore, future works will be directed to looking at the application of semi-supervised deep learning approaches to help improve prediction for reservoir properties, and facies analysis and also look at improving the DL model with some metaheuristic algorithms. Figure 10.6 shows a framework for the proposed semi-supervised deep learning model. In the semi-supervised learning algorithm architecture (Fig. 10.6), a small amount of labelled data is used for partial training of the model (supervised learning). After the supervised training, the trained model and the unlabelled dataset are used to predict the output which produces a pseudo label. Since these pseudo results may not be accurate, the algorithm links the pseudo labels data and the labelled training data at the same time linking the input labelled training data and that of the unlabelled training data. This linked data again uses the data to train the model to reduce the error and improve the accuracy of the model.

9.2 Rock Physics Perspective

Geophysical and petrophysical well logs are subjected to noise and various processing uncertainties, henceforth, a proper pre-processing technique is required to improve the qualitative and quantitative analysis of the data. Saikia et al. (2020) highlighted noise, outliers, spurious attribute removal, and a suitable correlation of different data sources in terms of time, depth, and frequency as some aspects of pre-processing technique. To handle these effectively, the application of rock physics has gained much attention in the society of geophysics. This paradigm has contributed to well log conditions. Also, rock physics in recent years has gained much attention in the oil industry to help condition the input well logs in other to improve reservoir characterisation (Chi and Han, 2009; Mukerji et al., 2001; Saberi, 2013; Saxena et al., 2018). To capture the heterogeneity and complexity of the rock's log response and the uncertainty associated with theoretical relations, a

Petroleum Reservoir Characterisation 251

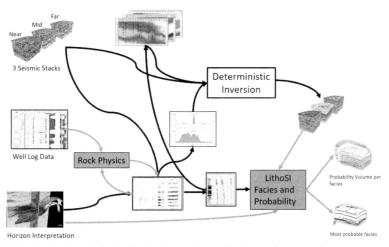

Fig. 10.7. Application of Rock Physics for well conditioning and its contribution to reservoir characterisation.

seismic-guided approach coupled with rock physics has been the best method. Since geological location has a great influence on the acquired data (geophysical, core data, and petrophysical), the characteristics (elastic moduli) of the geology can be addressed by the use of rock physics models and to understand the geophysical data response by such geology (Fig. 10.7). In Fig. 10.7, rock physics models are used to condition the various input logs (such as density and sonic logs) which are then used to better enhance reservoir characterisation (deterministic inversion, facies probability mapping, and others). The use of this technique helps in understanding the geological complexity, improving the well log response, and a considerable level of accuracy with a smaller number of data samples can be achieved. Therefore, future works should exploit the use of rock physics models alongside AI models to condition the various inputs used for the prediction of reservoir properties to have a more accurate and precise result for future property modelling.

10. Conclusions

The past years have seen the petroleum industry generate abundant data for its resource estimation. However, under current economic conditions, most field data acquisition has been limited thereby relying on the acquired data for reservoir characterisation taking into account the use of complex models. Integration of all available data (petrophysics, geology, geophysics, rock mechanics, and engineering) play an important role in achieving a good reservoir characterisation. The integration of the available data provides a high lateral resolution (seismic data), and high vertical resolution (well logs) while core data provides a physical measurement of the reservoir. Also, integrating these data sources helps to improve inter-well estimation taking into account the seismic data. Reservoir characterisation had been challenged by the fact that the various models have to handle complex nonlinearity present in the available dataset and also the various uncertainty present in the dataset and models. These flaws are mitigated by using strong data condition methods for dimensionality reduction, denoising, and identifying appropriate features for prediction to achieve accurate and reliable reservoir characterisation. The conclusion for the reservoir characterisation is highlighted as follows:

1. Most ML paradigms employed for petrophysical seismic characterisation have been reviewed. These ANN models have been used to improve the issues facing the empirical approach used to determine petrophysical parameters. Most literature highlights the use of SVM over ANN models since the former gives the least error, handles small datasets, and has faster processing times.

2. There has been an increase in the use of hybridizsd models in the field of reservoir characterisation. The hybrid models have helped to solve various problems facing conventional AI algorithms such as parameter optimisation, weight adjustment, computational time, uncertainty in data, and feature selections.

3. The selection of most hyperparameters for AI models' application to reservoir characterisation in the literature entails a guide technique of trial and error which is tedious and time-ingesting and often ends in suboptimal trained models. Hence, advanced studies must discover greater green strategies for using metaheuristics algorithms for the automatic selection of optimal hyperparameters and for improving computational time.

4. In a complex application, deep neural networks outperform traditional neural networks in handling high data complexity with automatic feature extraction. As a result, this method avoids the need for a separate feature extraction and selection phase.

5. Finally, while deep neural networks and hybridised metaheuristic models are widely used in many classification problems, their application has not been employed much in predicting reservoir properties. Seismic-driven integration model with deep learning and hybrid metaheuristic models with DL is yet to be explored in depth.

6. While data condition improves the overall prediction of various AI algorithms, the use of rock physics to help condition the various log inputs in this domain is yet to be explored in depth.

References

Al-Anazi, A. and Gates, I.D. 2010. A support vector machine algorithm to classify lithofacies and model permeability in heterogeneous reservoirs. *Engineering Geology* 114(3-4): 267–277. https://doi.org/10.1016/j.enggeo.2010.05.005.

Al-Anazi, A.F. and Gates, I.D. 2012. Support vector regression to predict porosity and permeability: Effect of sample size. *Computers and Geosciences* 39: 64–76. https://doi.org/10.1016/j.cageo.2011.06.011.

Al-Bulushi, N., King, P.R., Blunt, M.J. and Kraaijveld, M. 2009. Development of artificial neural network models for predicting water saturation and fluid distribution. *Journal of Petroleum Science and Engineering* 68(3–4): 197–208. https://doi.org/10.1016/j.petrol.2009.06.017.

Al-Mudhafar, W.J. 2017a. Integrating well log interpretations for lithofacies classification and permeability modeling through advanced machine learning algorithms. *Journal of Petroleum Exploration and Production Technology* 7(4): 1023–1033. https://doi.org/10.1007/s13202-017-0360-0.

Al-Mudhafar, W.J. 2017b. Integrating kernel support vector machines for efficient rock facies classification in the main pay of Zubair formation in South Rumaila oil field, Iraq. *Modeling Earth Systems and Environment* 3: 1–8. https://doi.org/10.1007/s40808-017-0277-0.

Alzubaidi, F., Mostaghimi, P., Swietojanski, P., Clark, S.R. and Armstrong, R.T. 2021. Automated lithology classification from drill core images using convolutional neural networks. *Journal of Petroleum Science and Engineering* 197: 107933. https://doi.org/10.1016/j.petrol.2020.107933.

Ameur-Zaimeche, O., Zeddouri, A., Heddam, S. and Kechiched, R. 2020. Lithofacies prediction in non-cored wells from the Sif Fatima oil field (Berkine basin, southern Algeria): a comparative study of multilayer perceptron neural network and cluster analysis-based approaches. *Journal of African Earth Sciences* 166: 103826. https://doi.org/10.1016/j.jafrearsci.2020.103826.

Amiri, M., Ghiasi-Freez, J., Golkar, B. and Hatampour, A. 2015. Improving water saturation estimation in a tight shaly sandstone reservoir using artificial neural network optimised by imperialist competitive algorithm: A case study. *Journal of Petroleum Science and Engineering* 127: 347–358. https://doi.org/10.1016/j.petrol.2015.01.013.

Anifowose, F. and Abdulraheem, A. 2011. Fuzzy logic-driven and SVM-driven hybrid computational intelligence models applied to oil and gas reservoir characterization. *Journal of Natural Gas Science and Engineering* 3(3): 505–517. https://doi.org/10.1016/j.jngse.2011.05.002.

Anifowose, F.A., Abdulraheem, A., Al-Shuhail, A. and Schmitt, D.P. 2013. Improved permeability prediction from seismic and log data using artificial intelligence techniques. pp. 2190–2196. *In: SPE Middle East Oil and Gas Show and Conference*. March. OnePetro. https://doi.org/10.2118/164465-MS.

Anifowose, F.A., Labadin, J. and Abdulraheem, A. 2015. Ensemble model of non-linear feature selection-based extreme learning machine for improved natural gas reservoir characterisation. *Journal of Natural Gas Science and Engineering* 26: 1561–1572. https://doi.org/10.1016/j.jngse.2015.02.012.

Anifowose, F.A., Labadin, J. and Abdulraheem, A. 2017. Hybrid intelligent systems in petroleum reservoir characterisation and modelling: The journey so far and the challenges ahead. *Journal of Petroleum Exploration and Production Technology* 7: 251–263. https://doi.org/10.1007/s13202-016-0257-3.

Archie, G.E. 1952. Classification of carbonate reservoir rocks and petrophysical considerations. *Aapg Bulletin* 36(2): 278–298. https://doi.org/10.1306/3D9343F7-16B1-11D7-8645000102C1865D.

Asante-Okyere, S., Shen, C., Ziggah, Y.Y., Rulegeya, M.M. and Zhu, X. 2020a. Principal component analysis (PCA) based hybrid models for the accurate estimation of reservoir water saturation. *Computers and Geosciences* 145: 104555. https://doi.org/10.1016/j.cageo.2020.104555.

Asante-Okyere, S., Shen, C., Ziggah, Y.Y., Rulegeya, M.M. and Zhu, X. 2020b. A novel hybrid technique of integrating gradient-boosted machine and clustering algorithms for lithology classification. *Natural Resources Research* 29: 2257–2273. https://doi.org/10.1007/s11053-019-09576-4.

Avseth, P. and Mukerji, T. 2002. Seismic lithofacies classification from well logs using statistical rock physics. *Petrophysics: The SPWLA Journal of Formation Evaluation and Reservoir Description* 43(02): 70–81.

Bae, H.S., Lee, H.J. and Lee, S.G. 2016. Voice recognition based on adaptive MFCC and deep learning. pp. 1542–1546. *In*: *2016 IEEE 11th Conference on Industrial Electronics and Applications (ICIEA)*, June. IEEE. 10.1109/ICIEA.2016.7603830.

Bestagini, P., Lipari, V. and Tubaro, S. 2017. A machine learning approach to facies classification using well logs. pp. 2137–2142. *In*: *Seg. Technical Program Expanded Abstracts 2017.* Society of Exploration Geophysicists. https://doi.org/10.1190/segam2017-17729805.1.

Bian, H., Xia, Y., Lu, C., Qin, X., Meng, Q. and Lu, H. 2020. Pore structure fractal characterization and permeability simulation of natural gas hydrate reservoir based on CT images. *Geofluids* 2020: 1–9. https://doi.org/10.1155/2020/6934691.

Bourbié, T., Coussy, O. and Zinszner, B. 1987. *Acoustics of Porous Media*. Gulf Publ. Co., translated by N. Marshall from French, *Acoustique des Milieux Poreu*. https://doi.org/10.1121/1.402899.

Bowers, M.C. and Fitz, D.E. 2000. A probabilistic approach to determine uncertainty in calculated water saturation. *In*: *SPWLA 41st Annual Logging Symposium*. June. OnePetro.

Brantson, E.T., Sibil, S., Osei, H., Owusu, E.B., Takyi, B. and Ansah, E. 2022. A new approach for saturation height modelling in a clastic reservoir using response surface methodology and artificial neural network. *Upstream Oil and Gas Technology* 9: 100081. https://doi.org/10.1016/j.upstre.2022.100081.

Carman, P.C. 1937. Fluid flow through granular beds. *Trans. Inst. Chem. Eng*. 15: 150–166. https://doi.org/10.1016/S0263-8762(97)80003-2.

Carman, P.C. 1956. Flow of gases through porous media. *Butterworths*, London. https://doi.org/10.1016/S0263-8762(97)80003-2.

Chai, X., Nie, W., Lin, K., Tang, G., Yang, T., Yu, J. and Cao, W. 2022. An open-source package for deep-learning-based seismic facies classification: benchmarking experiments on the SEG 2020 Open Data. *IEEE Transactions on Geoscience and Remote Sensing* 60: 1–19. 10.1109/TGRS.2022.3144666.

Chen, H., Engkvist, O., Wang, Y., Olivecrona, M. and Blaschke, T. 2018. The rise of deep learning in drug discovery. *Drug Discovery Today* 23(6): 1241–1250. https://doi.org/10.1016/j.drudis.2018.01.039.

Chi, X.G. and Han, D.H. 2009. Lithology and fluid differentiation using a rock physics template. *The Leading Edge* 28(1): 60–65. https://doi.org/10.1190/1.3064147.

Costa, A. 2006. Permeability-porosity relationship: A re-examination of the Kozeny-Carman equation based on a fractal pore-space geometry assumption. *Geophysical Research Letters* 33(2). https://doi.org/10.1029/2005GL025134.

De Matos, M.C., Osorio, P.L. and Johann, P.R. 2007. Unsupervised seismic facies analysis using wavelet transform and self-organising maps. *Geophysics* 72(1): 9–21. https://doi.org/10.1190/1.2392789.

Dong, S., Xu, L., Dai, Z., Xu, B.I.N., Yu, Q., Yin, S., Zhang, X., Zhang, C., Zang, X., Zhou, X. and Zhang, Z. 2020. A novel fractal model for estimating permeability in

low-permeable sandstone reservoirs. *Fractals* 28(08): 2040005. https://doi.org/10.1142/S0218348X20400058.
Dorrington, K.P. and Link, C.A. 2004. Genetic-algorithm/neural-network approach to seismic attribute selection for well-log prediction. *Geophysics* 69(1): 212–221. https://doi.org/10.1190/1.1649389.
Eaton, D., Caffagni, E., Van der Baan, M. and Matthews, L. 2014. Passive seismic monitoring and integrated geomechanical analysis of a tight-sand reservoir during hydraulic-fracture treatment, flowback, and production. pp. 1537–1545. *In*: *Unconventional Resources Technology Conference, Denver, Colorado, 25–27 August.* Society of Exploration Geophysicists, American Association of Petroleum Geologists, Society of Petroleum Engineers. https://doi.org/10.15530/urtec-2014-1929223.
Elshafei, M. and Hamada, G.M. 2009. Neural network identification of hydrocarbon potential of shaly sand reservoirs. *Petroleum Science and Technology* 27(1): 72–82. https://doi.org/10.2118/110959-MS.
Fang, S.H., Tsao, Y., Hsiao, M.J., Chen, J.Y., Lai, Y.H., Lin, F.C. and Wang, C.T. 2019. Detection of pathological voice using cepstrum vectors: A deep learning approach. *Journal of Voice* 33(5): 634–641. https://doi.org/10.1016/j.jvoice.2018.02.003.
Fujiyoshi, H., Hirakawa, T. and Yamashita, T. 2019. Deep learning-based image recognition for autonomous driving. *IATSS Research* 43(4): 244–252. https://doi.org/10.1016/j.iatssr.2019.11.008.
Furey, T.S., Cristianini, N., Duffy, N., Bednarski, D.W., Schummer, M. and Haussler, D. 2000. Support vector machine classification and validation of cancer tissue samples using microarray expression data. *Bioinformatics* 16(10): 906–914. https://doi.org/10.1093/bioinformatics/16.10.906.
Gaymard, R. and Poupon, A. 1968. Response of neutron and formation density logs in hydrocarbon bearing formations. *The Log Analyst* 9(05).
Gholami, R., Shahraki, A.R. and Jamali Paghaleh, M. 2012. Prediction of hydrocarbon reservoirs permeability using support vector machine. *Mathematical Problems in Engineering* 2012: 1–18. https://doi.org/10.1155/2012/670723.
Gogoi, T. and Chatterjee, R. 2019. Estimation of petrophysical parameters using seismic inversion and neural network modelling in Upper Assam basin, India. *Geoscience Frontiers* 10(3): 1113–1124. https://doi.org/10.1016/j.gsf.2018.07.002.
Hamada, G.M. and Elshafei, M.A. 2010. Neural network prediction of porosity and permeability of heterogeneous gas sand reservoirs using NMR and conventional logs. *Nafta* 61(10): 451–465. https://doi.org/10.2118/126042-MS.
Harrison, B. and Jing, X.D. 2001. Saturation height methods and their impact on volumetric hydrocarbon in place estimates. *In*: *SPE Annual Technical Conference and Exhibition.* September. OnePetro. https://doi.org/10.2118/71326-MS.
He, K., Zhang, X., Ren, S. and Sun, J. 2016. Deep residual learning for image recognition. pp. 770–778. *In*: *Proceedings of the IEEE Conference on Computer Vision and Pattern Recognition.* IEEE.
Heiba, A.A., Sahimi, M., Scriven, L.E. and Davis, H.T. 1992. Percolation theory of two-phase relative permeability. *SPE Reservoir Engineering* 7(01): 123–132. https://doi.org/10.2118/11015-PA.
Helle, H.B., Bhatt, A. and Ursin, B. 2001. Porosity and permeability prediction from wireline logs using artificial neural networks: A North Sea case study. *Geophysical Prospecting* 49(4): 431–444. https://doi.org/10.1046/j.1365-2478.2001.00271.x.
Helle, H.B. and Bhatt, A. 2002. Fluid saturation from well logs using committee neural networks. *Petroleum Geoscience* 8(2): 109–118. https://doi.org/10.1144/petgeo.8.2.109.

Hewett, T.A. 1986. Fractal distributions of reservoir heterogeneity and their influence on fluid transport. *In*: *SPE Annual Technical Conference and Exhibition*. October. OnePetro. https://doi.org/10.2118/15386-MS.

Hewett, T.A. and Behrens, R.A. 1990. Conditional simulation of reservoir heterogeneity with fractals. *SPE Formation Evaluation* 5(03): 217–225. https://doi.org/10.2118/18326-PA.

Hu, Y., Yu, X. and Chen, G. 2012. Classification of the average capillary pressure function and its application in calculating fluid saturation. *Petroleum Exploration and Development* 39(6): 778–784. https://doi.org/10.1016/S1876-3804(12)60104-9.

Huang, Z., Shimeld, J., Williamson, M. and Katsube, J. 1996. Permeability prediction with artificial neural network modelling in the Venture gas field, offshore eastern Canada. *Geophysics* 61(2): 422–436. https://doi.org/10.1190/1.1443970.

Huang, Z. and Williamson, M.A. 1997. Determination of porosity and permeability in reservoir intervals by artificial neural network modelling, offshore Eastern Canada. *Petroleum Geoscience* 3(3): 245–258. https://doi.org/10.1144/petgeo.3.3.245.

Iturrarán-Viveros, U. and Muñoz-García, M.A. 2018. Porosity and water saturation in sands or shales using Artificial Neural Networks and seismic attributes in a clastic reservoir in Colombia. pp. 1282–1285. *In*: *International Geophysical Conference, Beijing, China, 24–27 April 2018*. December. Society of Exploration Geophysicists and Chinese Petroleum Society. https://doi.org/10.1190/IGC2018-314.

Kaydani, H., Mohebbi, A. and Baghaie, A. 2012. Neural fuzzy system development for the prediction of permeability from wireline data based on fuzzy clustering. *Petroleum Science and Technology* 30(19): 2036–2045. https://doi.org/10.1080/10916466.2010.531345.

Kozeny, J. 1927. Uber kapillare Leitung des Wassers im Boden-Aufstieg, Versickerung und Anwendung auf die Bewasserung, Sitzungsberichte der Akademie der Wissenschaften Wien. *Mathematisch Naturwissenschaftliche Abteilung* 136: 271–306.

Li, D. and Lake, L.W. 1995. Scaling fluid flow through heterogeneous permeable media. *SPE Advanced Technology Series* 3(01): 188–197. https://doi.org/10.2118/26648-PA.

Lian, P.Q., Tan, X.Q., Ma, C.Y., Feng, R.Q. and Gao, H.M. 2016. Saturation modelling in a carbonate reservoir using capillary pressure based saturation height function: a case study of the Svk reservoir in the Y Field. *Journal of Petroleum Exploration and Production Technology* 6(1): 73–84. https://doi.org/10.1007/s13202-015-0159-9.

Lim, J.S. and Kim, J. 2004. Reservoir porosity and permeability estimation from well logs using fuzzy logic and neural networks. *In*: *SPE Asia Pacific Oil and Gas Conference and Exhibition*. OnePetro. https://doi.org/10.2118/88476-MS.

Luo, H., Tang, D. and Tang, Y. 2015. Study on prediction of oil water contact in carbonate reservoir with capillary pressure data. *Editorial Department of Petroleum Geology and Recovery Efficiency* 20(2): 71–73.

Ma, Y.Z. 2011. Uncertainty analysis in reservoir characterization and management: How much should we know about what we don't know? *In*: Ma, Y.Z. and P.R. La Pointe (eds.). *Uncertainty Analysis and Reservoir Modelling, AAPG Memoir*. 96: 1–15. DOI:10.1306/13301404M963458.

Mavko, G. and Nur, A. 1997. The effect of a percolation threshold in the Kozeny-Carman relation. *Geophysics* 62(5): 1480–1482. https://doi.org/10.1190/1.1444251.

McCulloch, W.S. and Walter, P. 1943. A logical calculus of the ideas immanent in nervous activity. *The Bulletin of Mathematical Biophysics* 5: 115–133. https://doi.org/10.1007/BF02478259.

Mehana, M. and El-monier, I. 2016. Shale characteristics impact on Nuclear Magnetic Resonance (NMR) fluid typing methods and correlations. *Petroleum* 2(2): 138–147. https://doi.org/10.1016/j.petlm.2016.02.002.

Meshri, I.D. 1986. On the reactivity of carbonic and organic acids and generation of secondary porosity: Roles of organic matter in sediment diagenesis. *In*: Gautier, D.L. (ed.). Roles of Organic Matter in Sediment Diagenesis. *SPEM Special Publication*. 38: 123–128.

Miller, R.S., Rhodes, S., Khosla, D. and Nino, F. 2019. Application of artificial intelligence for depositional facies recognition-permian basin. *In*: *Unconventional Resources Technology Conference*, Denver, Colorado, 22–24 July, Society of Exploration Geophysicists: pp. 4410–4415. https://doi.org/10.15530/urtec-2019-193.

Moghadasi, L., Ranaee, E., Inzoli, F. and Guadagnini, A. 2017. Petrophysical well log analysis through intelligent methods. *In*: *SPE Bergen One Day Seminar*. April. OnePetro. https://doi.org/10.2118/185922-MS.

Mohaghegh, S., Arefi, R., Ameri, S., Aminiand, K. and Nutter, R. 1996. Petroleum reservoir characterisation with the aid of artificial neural networks. *Journal of Petroleum Science and Engineering* 16(4): 263–274. https://doi.org/10.1016/S0920-4105(96)00028-9.

Mohamed, E., Elsayed, M., Hassan, A., Mahmoud, M. and El-Husseiny, A. 2022. A machine learning approach to predict the permeability from nmr t2 relaxation time distribution for various reservoir rock types. *In*: *ADIPEC*. October. OnePetro. https://doi.org/10.2118/211624-MS.

Mukerji, T., Avseth, P., Mavko, G., Takahashi, I. and González, E.F. 2001. Statistical rock physics: Combining rock physics, information theory, and geostatistics to reduce uncertainty in seismic reservoir characterization. *The Leading Edge* 20(3): 313–319. https://doi.org/10.1190/1.1438938.

Nelson, P.H. 1994. Permeability-porosity relationships in sedimentary rocks. *The Log Analyst* 35(03).

Nikravesh, M. and Aminzadeh, F. 2001. Past, present and future intelligent reservoir characterisation trends. *Journal of Petroleum Science and Engineering* 31(2–4): 67–79. https://doi.org/10.1016/S0920-4105(01)00121-8.

Nikravesh, M., Zadeh, L.A. and Aminzadeh, F. 2003. *Soft Computing and Intelligent Data Analysis in Oil Exploration*. Elsevier.

Nyein, C.Y., Ghareb, M., Hamada and Ahmed Elsakka. 2018. Artificial Neural Network (ANN) prediction of porosity and water saturation of shaly sandstone reservoirs. AAPG Asia Pacific Region, the 4th AAPG/EAGE/MGS Myanmar Oil and Gas Conference: Myanmar: A Global Oil and Gas Hotspot: Unleashing the Petroleum Systems Potential.

Okon, A.N., Adewole, S.E. and Uguma, E.M. 2021. Artificial neural network model for reservoir petrophysical properties: Porosity, permeability and water saturation prediction. *Modeling Earth Systems and Environment* 7(4): 2373–2390. https://doi.org/10.1007/s40808-020-01012-4.

Olakunle, I., Chinedu, A., Udoka, N. and Muyiwa, E. 2015. Saturation height modelling in a partially appraised gas field using analogue field core data: An optimisation case study of ZAN field in the Niger Delta. *In*: *SPE Nigeria Annual International Conference and Exhibition*. OnePetro. https://doi.org/10.2118/178374-MS.

Olson, T.M. 1998. Porosity and permeability prediction in low-permeability gas reservoirs from well logs using neura networks. *In*: *SPE Rocky Mountain Regional/Low-Permeability Reservoirs Symposium*. April. OnePetro. https://doi.org/10.2118/39964-MS.

Otchere, D.A., Ganat, T.O.A., Gholami, R. and Ridha, S. 2021a. Application of supervised machine learning paradigms in the prediction of petroleum reservoir properties: Comparative analysis of ANN and SVM models. *Journal of Petroleum Science and Engineering* 200: 108182. https://doi.org/10.1016/j.petrol.2020.108182.

Otchere, D.A., Ganat, T.O.A., Gholami, R. and Lawal, M. 2021b. A novel custom ensemble learning model for an improved reservoir permeability and water saturation prediction.

Journal of Natural Gas Science and Engineering 91: 103962. https://doi.org/10.1016/j.jngse.2021.103962.

Otchere, D.A., Abdalla Ayoub Mohammed, M., Ganat, T.O.A., Gholami, R. and Aljunid Merican, Z.M. 2022a. A novel empirical and deep ensemble super learning approach in predicting reservoir wettability via well logs. *Applied Sciences* 12(6): 2942. https://doi.org/10.3390/app12062942.

Otchere, D.A., Ganat, T.O.A., Nta, V., Brantson, E.T. and Sharma, T. 2022b. Data analytics and Bayesian Optimised Extreme Gradient Boosting approach to estimate cut-offs from wireline logs for net reservoir and pay classification. *Applied Soft Computing* 120: 108680. https://doi.org/10.1016/j.asoc.2022.108680.

Ouenes, A. 2000. Practical application of fuzzy logic and neural networks to fractured reservoir characterization. *Computers & Geosciences* 26(8): 953–962. https://doi.org/10.1016/S0098-3004(00)00031-5.

Pettijohn, F.J. 1975. *Sedimentary Rocks*. New York: Harper & Row. Vol. 3, p. 628.

Pouyanfar, S., Sadiq, S., Yan, Y., Tian, H., Tao, Y., Reyes, M.P., Shyu, M.L., Chen, S.C. and Iyengar, S.S. 2018. A survey on deep learning: Algorithms, techniques, and applications. *ACM Computing Surveys (CSUR)* 51(5): 1–36. https://doi.org/10.1145/3234150.

Qiang, Z., Yasin, Q., Golsanami, N. and Du, Q. 2020. Prediction of reservoir quality from log-core and seismic inversion analysis with an artificial neural network: A case study from the Sawan Gas Field, Pakistan. *Energies* 13(2): 486. https://doi.org/10.3390/en13020486.

Qiu, Y.N. 2000. Development of geological reservoir modelling in past decade, Beijing (in Chinese). *Acta Petrolei Sinica*. 12(4): 55–62.

Rafiq, A., Eaton, D.W., McDougall, A. and Pedersen, P.K. 2016. Reservoir characterization using microseismic facies analysis integrated with surface seismic attributes. *Interpretation* 4(2): 167–181. https://doi.org/10.1190/INT-2015-0109.1.

Rere, L.M., Fanany, M.I. and Arymurthy, A.M. 2016. Metaheuristic algorithms for convolution neural network. *Computational Intelligence and Neuroscience*. https://doi.org/10.1155/2016/1537325.

Reza, C., Mohammed, A.R. and Mahmund, M. 2017. Determination of the petrophysical parameters using geostatistical method in one of the hydrocarbon reservoirs in South West of Iran. *J. Sci. Eng. Res.* 4(12): 44–55.

Rudyk, S. and Al-Lamki, A. 2015. Saturation-height model of omani deep tight gas reservoir. *Journal of Natural Gas Science and Engineering* 27: 1821–1833. https://doi.org/10.1016/j.jngse.2015.11.015.

Saberi, M.R. 2013. Rock physics integration: From petrophysics to simulation. *In*: *10th Biennal International Conference and Expositions.* Vol. 444.

Saemi, M., Ahmadi, M. and Varjani, A.Y. 2007. Design of neural networks using genetic algorithm for the permeability estimation of the reservoir. *Journal of Petroleum Science and Engineering* 59(1–2): 97–105. https://doi.org/10.1016/j.petrol.2007.03.007.

Sahimi, M. 1993. Flow phenomena in rocks: From continuum models to fractals, percolation, cellular automata, and simulated annealing. *Reviews of Modern Physics* 65(4): 1393. https://doi.org/10.1103/RevModPhys.65.1393.

Saikia, P., Baruah, R.D., Singh, S.K. and Chaudhuri, P.K. 2020. Artificial Neural Networks in the domain of reservoir characterisation: A review from shallow to deep models. *Computers and Geosciences* 135: 104357. https://doi.org/10.1016/j.cageo.2019.104357.

Saxena, V., Krief, M. and Adam, L. 2018. *Handbook of Borehole Acoustics and Rock Physics for Reservoir Characterization*. Elsevier.

Schwartz, L.M. and Kimminau, S. 1987. Analysis of electrical conduction in the grain consolidation model. *Geophysics* 52(10): 1402–1411. https://doi.org/10.1190/1.1442252.

Sebtosheikh, M.A. and Salehi, A. 2015. Lithology prediction by support vector classifiers using inverted seismic attributes data and petrophysical logs as a new approach and investigation of training data set size effect on its performance in a heterogeneous carbonate reservoir. *Journal of Petroleum Science and Engineering* 134: 143–149. https://doi.org/10.1016/j.petrol.2015.08.001.

Shaheen, F., Verma, B. and Asafuddoula, M. 2016. Impact of automatic feature extraction in deep learning architecture. pp. 1–8. *In*: *2016 International Conference on Digital Image Computing: Techniques and Applications (DICTA)*. IEEE. 10.1109/DICTA.2016.7797053.

Shedid, S.A. and Saad, M.A. 2017. Comparison and sensitivity analysis of water saturation models in shaly sandstone reservoirs using well logging data. *Journal of Petroleum Science and Engineering* 156: 536–545. https://doi.org/10.1016/j.petrol.2017.06.005.

Shen, C., Asante-Okyere, S., Yevenyo Ziggah, Y., Wang, L. and Zhu, X. 2019. Group method of data handling (GMDH) lithology identification based on wavelet analysis and dimensionality reduction as well log data pre-processing techniques. *Energies* 12(8): 1509. https://doi.org/10.3390/en12081509.

Shook, M., Li, D. and Lake, L.W. 1992. Scaling immiscible flow through permeable media by inspectional analysis. *In Situ (United States)* 16(4).

Singh, S., Kanli, A.I. and Sevgen, S. 2016. A general approach for porosity estimation using artificial neural network method: A case study from Kansas gas field. *Studia Geophysica et Geodaetica* 60: 130–140. https://doi.org/10.1007/s11200-015-0820-2.

Soleimani, F., Hosseini, E. and Hajivand, F. 2020. Estimation of reservoir porosity using analysis of seismic attributes in an Iranian oil field. *Journal of Petroleum Exploration and Production Technology* 10(4): 1289–1316. https://doi.org/10.1007/s13202-020-00833-4.

Srisutthiyakorn, N. 2016. Deep-learning methods for predicting permeability from 2D/3D binary-segmented images. pp. 3042–3046. *In*: *SEG Technical Program Expanded Abstracts 2016*). Society of Exploration Geophysicists. https://doi.org/10.1190/segam2016-13972613.1.

Tang, H. 2008. Improved carbonate reservoir facies classification using artificial neural network method. *In*: *Canadian International Petroleum Conference*. June. OnePetro. https://doi.org/10.2118/2008-122.

Timur, A. 1968. An investigation of permeability, porosity, and residual water saturation relationships. *In*: *SPWLA 9th Annual Logging Symposium*. June. OnePetro.

Tran, T.V., Ngo, H.H., Hoang, S.K., Tran, H.N. and Lambiase, J.J. 2020. Depositional facies prediction using artificial intelligence to improve reservoir characterisation in a mature field of Nam con son basin, offshore Vietnam. *In*: *Offshore Technology Conference Asia*. October. OnePetro. https://doi.org/10.4043/30086-MS.

Verma, A.K., Cheadle, B.A., Routray, A., Mohanty, W.K. and Mansinha, L. 2012. Porosity and permeability estimation using neural network approach from well log data. pp. 1–6. *In*: *SPE Annual Technical Conference and Exhibition* (May).

Walker, R.G. 1992. Facies, facies models, and modern stratigraphic concepts. *Facies Models: Response to Sea Level Change*: p. 434. https://doi.org/10.4116/jaqua.34.19.

Wang, D., Peng, J., Yu, Q., Chen, Y. and Yu, H. 2019. Support vector machine algorithm for automatically identifying depositional microfacies using well logs. *Sustainability* 11(7): 1919. https://doi.org/10.3390/su11071919.

Wong, K.W., Fung, C.C., Ong, Y.S. and Gedeon, T.D. 2005. Reservoir characterization using support vector machines. pp. 354–359. *In*: *International Conference on Computational Intelligence for Modelling, Control and Automation and International Conference on Intelligent Agents, Web Technologies and Internet Commerce (CIMCA-IAWTIC'06)*, November. *IEEE*. Vol. 2. 10.1109/CIMCA.2005.1631494.

Wyllie, M.R.J. and Rose, W.D. 1950. Some theoretical considerations related to the quantitative evaluation of the physical characteristics of reservoir rock from electrical log data. *Journal of Petroleum Technology* 2(04): 105–118. https://doi.org/10.2118/950105-G.

Xie, Y., Zhu, C., Zhou, W., Li, Z., Liu, X. and Tu, M. 2018. Evaluation of machine learning methods for formation lithology identification: A comparison of tuning processes and model performances. *Journal of Petroleum Science and Engineering* 160: 182–193. https://doi.org/10.1016/j.petrol.2017.10.028.

Xie, Y., Zhu, C., Hu, R. and Zhu, Z. 2021. A coarse-to-fine approach for intelligent logging lithology identification with extremely randomised trees. *Mathematical Geosciences* 53: 859–876. https://doi.org/10.1007/s11004-020-09885-y.

Xie, Y., Zhu, C., Hu, R. and Zhu, Z. 2021. A coarse-to-fine approach for intelligent logging lithology identification with extremely randomised trees. *Mathematical Geosciences* 53: 859–876. https://doi.org/10.1007/s11004-020-09885-y.

Yu, B. and Cheng, P. 2002. A fractal permeability model for bi-dispersed porous media. *International Journal of Heat and Mass Transfer* 45(14): 2983–2993. https://doi.org/10.1016/S0017-9310(02)00014-5.

Yu, X., Ma, Y.Z., Psaila, D. et al. 2011. Reservoir characterisation and modelling: A look back to see the way forward. *In*: Ma, Y.Z. and P.R. La Pointe (eds.). *Uncertainty Analysis and Reservoir Modelling, AAPG Memoir.* 96: 289–309. DOI:10.1306/13301421M963458.

Yu, X.H. 2008. *Hydrocarbon Reservoir Sedimentology of Clastic Sandstone* (2nd Edn.) (in Chinese): Beijing, China: Petroleum Industry Press: p. 551.

Yu, Y., Lin, L., Zhai, C., Chen, H., Wang, Y., Li, Y. and Deng, X. 2019. Impacts of lithologic characteristics and diagenesis on reservoir quality of the 4th member of the Upper Triassic Xujiahe Formation tight gas sandstones in the western Sichuan Basin, southwest China. *Marine and Petroleum Geology* 107: 1–19. https://doi.org/10.1016/j.marpetgeo.2019.04.040.

Zabihi, R., Schaffie, M., Nezamabadi-Pour, H. and Ranjbar, M. 2011. Artificial neural network for permeability damage prediction due to sulfate scaling. *Journal of Petroleum Science and Engineering* 78(3–4): 575–581. https://doi.org/10.1016/j.petrol.2011.08.007.

Zendehboudi, S., Rezaei, N. and Lohi, A. 2018. Applications of hybrid models in chemical, petroleum, and energy systems: A systematic review. *Applied Energy* 228: 2539–2566. https://doi.org/10.1016/j.apenergy.2018.06.051.

Zerrouki, A.A., Aifa, T. and Baddari, K. 2014. Prediction of natural fracture porosity from well log data by means of fuzzy ranking and an artificial neural network in Hassi Messaoud oil field, Algeria. *Journal of Petroleum Science and Engineering* 115: 78–89. https://doi.org/10.1016/j.petrol.2014.01.011.

Zheng, B. and Li, J.H. 2015. A new fractal permeability model for porous media based on Kozeny-Carman equation. *Natural Gas Geoscience* 26(1): 193–198. https://doi.org/10.1155/2022/8088151.

Zhong, Z. and Carr, T.R. 2019. Application of a new hybrid particle swarm optimization-mixed kernels function-based support vector machine model for reservoir porosity prediction: A case study in Jacksonburg-Stringtown oil field, West Virginia, USA. *Interpretation* 7(1): 97–112. https://doi.org/10.1190/INT-2018-0093.1.

CHAPTER 11

Artificial Lift Design for Future Inflow and Outflow Performance for Jubilee Oilfield
Using Historical Production Data and Artificial Neural Network Models

Solomon Adjei Marfo,[1] *Eric Thompson Brantson,*[2,*]
Eric Mensah Amarfio,[2] *Abakah-Paintsil Efua Eduamba,*[2]
Iyiola Zainab Ololade,[2] *Alexander Mensah Ofori,*[2]
Ebenezer Ansah[3] *and Emmanuel Karikari Duodu*[2]

1. Introduction

Forecasting production parameters such as production rate, oil recovery, or reserves estimation is an essential aspect that enables operators to determine the economic profitability of a petroleum venture. Inflow and Tubing Performance Relationships (IPR and TPR) are the two mathematical techniques used to analyse and predict the performance of a well. AL-Dogail et al. (2018) used artificial intelligence (AI) (backpropagation network and fuzzy logic) as another technique to predict the Inflow IPR of a gas reservoir for effective reservoir management.

---

[1] Department of Chemical and Petrochemical Engineering, GNPC School of Petroleum Studies, University of Mines and Technology, Tarkwa, Ghana.
[2] Department of Petroleum and Natural Gas Engineering, GNPC School of Petroleum Studies, University of Mines and Technology, Tarkwa, Ghana.
[3] Department of Petroleum Geosciences and Engineering, GNPC School of Petroleum Studies, University of Mines and Technology, Tarkwa, Ghana.
* Corresponding author: etbrantson@umat.edu.gh

The operator of Jubilee Field, Ghana, recently updated its production figures downward from 87,000 to 70,000 Barrel of Oil Per Day (BOPD) due to certain technical issues such as increasing water cuts (Renpu, 2011), reduced gas offtake by Atuabo Gas, and sand contamination of the flowlines among a host of others (Agyeman, 2020). Initially, field assessments made by operators of Jubilee Field, Tullow Oil, indicated a production peak rate of 120,000 BOPD, water injection capacity of 230,000 Barrel of Water Per Day (BWPD), and gas export and injection capacity of up to 160 MMscf/d (Schempf, 2011). As a result of these technical problems, this production rate has been cut down. These myriads of challenges can be solved by means of implementing an artificial lift system (Boyun et al., 2007).

Artificial lift generally is a means of lowering the bottomhole flowing pressure so that a well can produce at some desired rate. Traditionally, the artificial lift had been referred to as downhole pumping or gas lift activity associated with mature fields, where the average reservoir pressure has declined such that the reservoir can no longer produce from its own natural energy (Mohammed et al., 2016). About 90% of all producing wells require some form of artificial lift to increase the flow of fluids from the wells when the reservoir no longer has sufficient energy to naturally produce at economic rates for enhanced financial performance (Flatern, 2015). When reservoir energy is insufficient or water cut is higher despite sufficient energy, then the artificial lift method could be used to maintain oil well production under rational production pressure drawdown (Fleshman et al., 2011). Geographical and environmental circumstances were considered the dominant factors in the selection of artificial lifting and some other subordinated factors, including reservoir pressure, productivity index, properties of reservoir fluid, and inflow performance according to Neely et al. (1981). Some of these lift methods are suited for high-rate wells whiles others are useful for low-rate wells. Several factors are taken into consideration before an artificial lift method is selected for a specific application (Battia and McAllister, 2014; Woods and Lea, 2017).

To address these challenges and enhance production, this study aims to investigate the feasibility of screening and designing an optimum artificial lift method to increase production and predict future IPR and TPR for the Jubilee Field using historical production data. Also, this study investigated the use of Radial Basis Function Neural Network (RBFNN) and Backpropagation neural network (BPNN) models to predict Tubing Head Pressure (THP) using historical data. A similar application study was carried out by Wu et al. (2016) and Brantson et al. (2018) in forecasting tight gas carbonate reservoir production profiles. Furthermore, the application of Artificial Neural Networks (ANNs) in the oil and gas industry began in the 1980s, however, it has become rampant over this decade and is now a common industry practice (Baudoin, 2016; Brantson et al., 2019a, 2019b; Chen et al., 2021).

This paper is structured into the following sections. Section 2 dwells on the method used to forecast future IPR and TPR as well as the ANN prediction of THP. Section 3 states the results obtained from the implementation of the IPR, TPR, and ANN. Section 4 summarises the major conclusion obtained from the study.

2. Methodology

2.1 Artificial Lift Screening Techniques

To apply artificial lift techniques to a field, screening must be done to ensure that the right artificial lift system is applied to the field in question. Some factors to be considered during the screening process are location, depth, estimated production, reservoir properties, and other factors. The screening process is the initial procedure to evaluate the suitability of a certain artificial lifting system (Brown, 1982). The goal is to phase out inappropriate systems progressively, reducing the selection to a few contenders for the next selection process. It is attempted to choose the optimum system, and to compare design parameters with existing methodologies and charts (Lea and Nickens, 1999). In this study, field parameters were compared to the parameters on the screening chart (Takacs, 2015) and the appropriate lift method was selected.

2.2 Inflow Performance Relationship Production Forecast

An inflow performance relationship is a means of evaluating a reservoir's deliverability in oil and gas production. An IPR curve is a graphical representation of the relation between the flowing bottomhole pressure and liquid production rate (Boyun et al., 2007). IPR curves are presented in a standardised manner, with the flowing bottomhole pressure on the ordinate of the graph and the corresponding production rate on the abscissa (Hill et al., 1993).

Equations (11.1) and (11.2) are based on Vogel's model and they can be used to generate a family of curves. This is done to improve the future performance of the well, which can be utilised to assess the good performance throughout the whole life of a well by the design engineer (Boyun et al., 2007).

$$\frac{J_f^*}{J_p^*} = \frac{\left(\frac{k_{ro}}{B_o \mu_o}\right)_f}{\left(\frac{k_{ro}}{B_o \mu_o}\right)_p} \qquad (11.1)$$

$$q = \frac{J_f^* \bar{p}_f}{1.8}\left[1 - 0.2\frac{p_{wf}}{\bar{p}_f} - 0.8\left(\frac{p_{wf}}{\bar{p}_f}\right)^2\right] \quad (11.2)$$

where,

$\bar{p}_f$ = reservoir pressure in a future time.
J_f^* = future productivity index
J_p^* = present productivity index
k_{ro} = relative permeability to oil
B_o = oil formation volume factor
μ_o = oil viscosity
p_{wf} = bottomhole pressure
q = flowrate

2.3 Outflow Performance Relationship Production Forecast

Vertical lifting performance (VLP) also known as Outflow Performance Relationship (OPR) is a description of bottomhole pressure as a function of flow rate. Different correlations for VLP are provided for outflow performance analysis.

2.4 PROSPER Procedure for Well Model Set-Up

PROSPER (Production and System Performance) was used to build a separate model for each component of a well system that contributes to overall performance and then enables each model subsystem to be verified by performance matching. The program, therefore, guarantees the exactness of the calculation. When the model system is modified to real data, PROSPER is reliably employed to simulate different scenarios based on surface production data and to anticipate the reservoir pressure.

PROSPER has five main sections. They are the Options Summary, PVT Data, IPR Data, Equipment Data, and Analysis Summary. The following steps were used to build the model, perform sensitivity analysis, and design an optimum gas lift model.

Step 1: Options Summary

Well characteristics, fluid type, well completion, and desired lift method are defined in this section as shown in Table 11.1. The Black Oil model is selected since it is suitable for a variety of applications and hydrocarbon systems.

Table 11.1. Options summary data.

Data	Remarks
Options	The options menu is used to define the characteristics of the well. Well characteristics such as fluid type, well completion, and desired lift method are defined here.
PVT Model	The Black Oil model was selected. This model is suitable for usage in a wide variety of applications and hydrocarbon fluid systems. A minimum of GOR solution, oil viscosity and water formation salinity are required. Data from PVT are imputed, and correlations have been chosen that best corresponds to the location or oil type.
Fluid Description	Oil and Water
Temperature Model	Rough approximation
Flow Type	Tubing flow
Well Type	Producer
Prediction	Pressure and temperature offshore model
Well Completion	Cased hole
Reservoir Type	Single branch reservoir
Artificial Lift Design	Gas lift

Step 2: PVT Data

Data such as Solution Gas Oil Ratio (GOR), water salinity, and impurities present in the oil are imputed into the simulation program as shown in Table 11.2. Data for matching actual field data to the simulation program is also imputed in this menu. The bubble point and oil viscosity correlation were selected. Glazo and Beal et al. correlations were the best correlation models with the least standard deviation.

Table 11.2. PVT data.

Input Data	Values
Solution GOR (scf/bbl)	1 243
Oil Gravity (°API)	36
Gas Gravity	0.878
Water Salinity (ppm)	100000
Mole percent H_2S (%)	0
Mole percent CO_2 (%)	0
Mole percent N_2 (%)	0

Step 3: Equipment Data

Under this section of the PROSPER software, well parameters such as well deviation, surface equipment, downhole equipment, geothermal gradient, and average heat capacities are keyed in.

2.4.1 Deviation Survey Data Input

The measured depth of the reservoir and the true vertical depth were entered. Because all the wells under consideration are vertical, top, and bottom perforations were used. The data used can be seen in Table 11.3.

Table 11.3. Deviation survey data.

Parameters	Values
Datum at Christmas Tree (ft)	0
Measured Depth (ft)	12519
True Vertical Depth (ft)	12300

2.4.2 Surface Equipment Data Input

Information entered in this section includes the Christmas tree, manifold, or choke. An ambient temperature of 60°F and an overall heat transfer coefficient of 8 BTU/hr/ft^2/F were keyed in. Data is found in Table 11.4.

Table 11.4. Surface equipment data.

Parameters	Values
Manifold/Christmas Tree (True Vertical Depth)	0
Temperature of Surroundings (ft)	60 °F
Overall Heat Transfer	8 BTU/hr/ft^2/F

2.4.3 Downhole Equipment Data Input

This section allows the user to input the tubing and casing inside and outside diameters. Also, the roughness coefficient of casing and tubing can be imputed, otherwise, the theoretical value of 0.0006 can be used if the roughness is unknown. Table 11.5 shows downhole data.

Table 11.5. Downhole equipment data.

Equipment	Measured Depth (ft)	Tubing/Casing Inside Diameter (inches)	Tubing/Casing Inside Roughness (inches)
Tubing	12 300	4.778	0.0006
Casing	12 519	8.535	0.0006

2.4.4 Average Heat Capacities Data Input

This section allows the user to input the average heat capacities for gas, oil, and water. Table 11.6 shows the data imputed.

Table 11.6. Average heat capacities data.

Parameters	Values
Cp oil	0.53 BTU/lb/F
Cp gas	0.51 BTU/lb/F

Step 4: Sensitivity Analysis

PROSPER helps users perform analysis by changing parameters such as GOR, water cut, and reservoir pressure to know how changes in these parameters will affect the well's or reservoirs' deliverability. For this work, water cut, GOR, and reservoir pressure were varied.

Step 5: Vertical Lift Performance Correlations

Different correlations for VLP are provided. Test results showed that the Petroleum Expert correlation was appropriate to the well circumstances and therefore employed for VLP and sensitivity analysis. For surface equipment horizontal piping, Beggs and Brill (1991) correlations were employed.

Step 6: Gas Lift Modelling

For this work, a continuous gas lift system was employed. The reason for opting for the continuous gas lift was based on these criteria; wells in this work are high productivity wells, lift gas is available for use and this warrants the use of gas lift method and field under investigation produces under high GOR conditions. PROSPER simulation software was used in the modelling of the gas lift for all the wells under investigation. A gas injection rate was 10 MMscf/day was used. The casing pressure imputed was at 1,500 psi with a kickoff pressure of about 1,900 psi. The pressure gradient of the kill fluid was set to 0.465 psi/ft. All these datasets are shown in Table 11.7.

2.5 Artificial Neural Networks

ANN is a group of inputs and outputs of connected units where each connection has a weight associated with its computer program (Pennel et al., 2018; Akwensi et al., 2021; Brantson et al., 2022). This network helps build predictive models from large databases. The model is built on the concept of the human nervous system. For this work, two types of ANN (BPNN and RBFNN) models will be considered.

Table 11.7. Gas lift data.

Parameters	Values
Minimum valve spacing	250 ft
Kill fluid gradient	0.465 ft/psi
Operating Injection Pressure	1300 psi
Kick off pressure	1900 psi
Gas lift valve type	R-20 Monel Port Size: 32 R-value: 0.25
Differential pressure dP across valve	250 psi
Vertical Lift Performance correlation	Petroleum Experts 2
Surface Equipment correlation	Beggs and Brill
Well Depth	12300 ft
Water Cut	14%
Maximum Liquid rate	12472 STB/D
Maximum Gas Available	10 MMScf/day

2.5.1 Back Propagation Neural Network

BPNN is a type of ANN model that involves fine-tuning the weights of the neural network based on the error rate obtained in a previous iteration. Proper tuning of the weights allows one to reduce the error rate which makes the model more accurate thereby increasing the generalisation of the model.

2.5.2 Radial Basis Function Neural Network

RBFNN is a type of neural network which approximates multivariable functions or multivariate by a linear combination of terms so that it can be used in more than one dimension. It is applied to approximate functions or data which is known at a finite number of points so that evaluations of the approximate function can take place often and efficiently.

2.5.3 ANN Procedure

For this work, historical production data from the Jubilee field which is the production data from the wells was grouped into training and testing data. MATLAB was the software used in building the ANN models. The inputs for the prediction were days, gas rate, cumulative gas produced, cumulative oil production, and flowing bottomhole pressure. The output for the model was the THP. To test the accuracy of the model, some statistical parameters were measured as the MAPE and correlational coefficient (R). A value of R closer to 1 means good accuracy while a value closer to 0 for MAPE indicates the

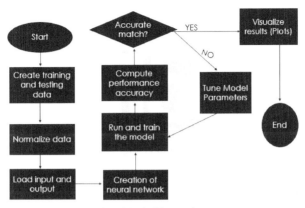

Fig. 11.1. ANN procedure.

least amount of error. Computational time was also calculated to know which model works faster. Figure 11.1 shows the process followed to obtain results from the ANN models.

3. Results and Discussion

3.1 Production and Well Data of the Study Area

The Jubilee Field is located in Western Region, 60 km offshore, between the Deepwater Tano and West Cape Three Points blocks, off the coast of Ghana. Figure 11.2 shows the Jubilee concession and payout, respectively. The field lies within the tropical climatic zone between latitude 4.492°N and longitude 2.916°W. The Jubilee Field appraisal and development programme began at the end of 2008 with the drilling of the Odum, Mahogany, and Heydua wells. The field's recoverable reserves are estimated to be more than 370 million barrels, with an upside potential of 1.8 billion barrels. It is located at a water depth of 1100 m. The Jubilee Field was developed through an FPSO (floating production storage and offloading system) with a target plateau oil rate of 120,000 BOPD, water injection capacity of 230,000 BWPD, and gas export and injection capacity of up to 160 MMscf/d (Schempf, 2011). Subsea gathering receptacles are located below the riser base to receive production from wells for onward transportation to an FPSO for processing. Compressors are installed on the FPSO to process produced gas for delivery to an onshore gas processing facility. Part of this processed gas is injected at the riser base for intermittent subsea boosting.

The obtained production data used consisted of daily gas production (MMSCF/D), BOPD, cumulative oil production (MMSTB), Tubing Head Pressure (psig), Flowing Bottom Hole Pressure (psia), and number of days of production.

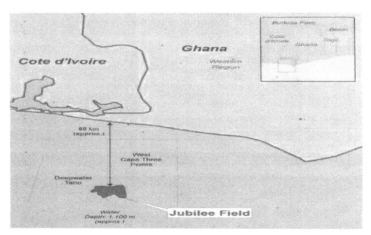

Fig. 11.2. Jubilee field concession.

Two random wells were chosen for evaluation. Obtained data conforms to American Petroleum Institute Recommended Practice API RP 11V6 for gas lift design criteria (Agyemang, 2020). Well data obtained constitutes Pressure Volume and Temperature (PVT) data, well inflow characteristics data, fluid properties data, wellbore geometry, surface facilities, and fluid properties data. For this work, the two randomly selected wells are named A and B. All the wells are vertical wells. Tables 11.8–11.11 are the well data obtained from Agyemang (2020).

Table 11.8. Production system data.

Parameters	Well A	Well B
Production Interval: Top: (ft) Bottom: (ft)	12300 12519	11559 11724
Casing Inside Diameter: in Weight: lb/ft	8.535 53.5	8.535 53.5
Tubing OD: in	5.5	5.5
Grade: ksi13Cr	80	80
Tubing ID: in Burst Pressure: psi	4.778 9190	4.778 9190
Weight (lb/ft)	20	20
PBTD: (ft)	12850	11832

(*Source*: Agyemang, 2020)

Table 11.9. Well surface data.

Parameters	Well A	Well B
Separator/Well Head Pressure (Psi)	441	441
Temperature (°F)	110	110
Primary Power (V)	415	415
Frequency (Hz)	60	60
Oil specific gravity (°API)	36	36
Water specific gravity	1.03	1.03
Scaling	nil	nil
Gas specific Gravity	0.878	0.878
CO_2 content H_2S content Paraffin Asphaltenes Sand production	nil " " " "	nil " " " "
Water cut (%)	14	20
GOR (scf/bbl)	1243	1243
Bubble point pressure: (psia)	4419	4419

Table 11.10. PVT and viscosity data.

Well	A	B
Reservoir Pressure (psia)	6 014.7	6 014.7
Temperature (°F)	210	210
Oil Formation Volume factor (Bo) (bbl/stb)	1.595	1.595
Solution Gas Oil Ratio (GOR) (scf/stb)	1243	1 43
Oil Viscosity μ_o (cP)	0.7	0.7

Table 11.11. Well inflow characteristics data.

Well Test (production)	A	B
Static Pressure, psig	5516	4872
Test Rate (bbl/day)	14008	3026
Test Pressure, psig	3701	4635

3.1.1 Base Case Flow Rates

The results below graphically display the calculated average daily flow rates of Well A and B. Table 11.12 shows the results from the calculation. This will serve as the basis for checking if there has been an improvement in wellbore deliverability after implementing the continuous gas lift method. Figures 11.3 and 11.4 show the daily production profile for well A and B with average base case flow rates of 8,161 BOPD and 9,850 BOPD, respectively.

Table 11.12. Screening results.

Field Parameters	Values
Depth	3657.6 m
Production Volume	24 000 bbl/day
Temperature	98.89–106.11°C
Oil Gravity	36°API
Prime Mover	Available Compressors
Location	Offshore

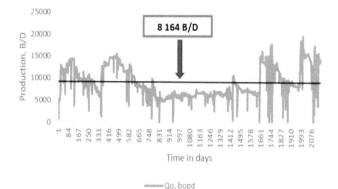

Fig. 11.3. Well A daily production profile.

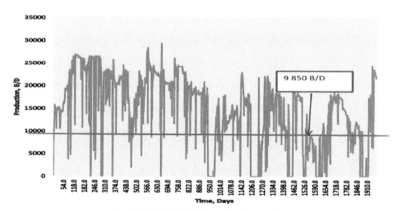

Fig. 11.4. Well B daily production profile.

3.2 Artificial Lift Screening

Table 11.12 shows the field parameters compared to the parameters on the chart by Takacs (2015) and the gas lift method was selected which satisfies the chart. The following are the results obtained during screening for an appropriate artificial lift method for the field under investigation in this work.

Since both wells are high productivity index wells, a screening chart for high productivity wells was used. Sucker rod and progressive cavity pumps were not applicable in this field due to the depth restriction of not more than 3,050 m. Electrical Submersible Pump (ESP) and Gas Lift were the next in line for the screening process. At the end of the screening process, the gas lift was selected based on the following criteria. Gas lift can handle fields that have wells with a fluid gravity greater than 15° API. Also, the temperature for all the wells was in the acceptable range for gas lift 98.89–106.11°C. The required operating volumes for the wells were within an acceptable range for gas lift use. The required operating volumes for the gas lift range is 200–30,000 BOPD. Based on screening by advantages, gas lift is an excellent choice for offshore applications due to its high efficiency rate when used offshore. In addition to that, compressors are already available onsite for use. For the above reasons, the gas lift was chosen as the artificial lift technique for the wells under investigation for this research. Table 11.12 shows the field data used for the screening.

3.3 PROSPER Simulation Results

3.3.1 IPR Curves

PROSPER simulation software was used to generate IPR family curves for each well. PROSPER software generates the curves as well as calculates the Absolute Open Flow (AOF) potential. Figures 11.5 and 11.7 show the current IPR curves for well A and B, respectively. The intercept on the ordinate axis

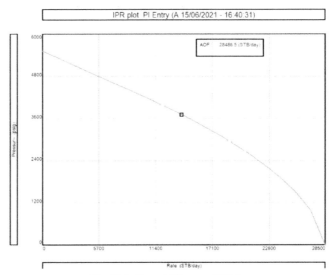

Fig. 11.5. Present IPR curve Well A.

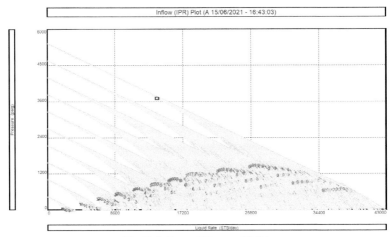

Fig. 11.6. Future IPR curve Well A.

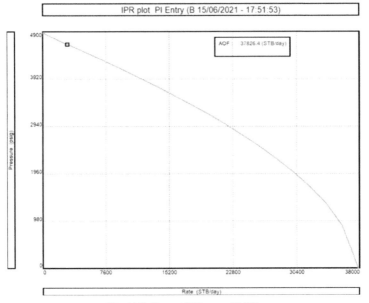

Fig. 11.7. Present IPR curve Well B.

indicates the shut-in reservoir pressure whereas the intercept on the abscissa indicates the flowrate in stock tank barrels per day. The AOF for Well A is 28,486.5 STB/D and that of Well B is 37,826.4 STB/D. Future IPR curves generated through sensitivity analysis can be seen in Fig. 11.6 and 11.8 for well A and B, respectively.

Artificial Lift Design for Future Inflow and Outflow Performance for Jubilee Oilfield 275

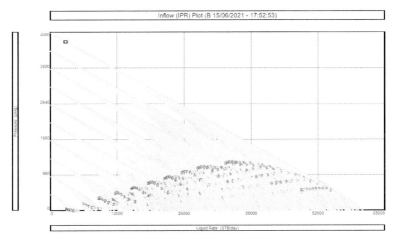

Fig. 11.8. Future IPR curve Well B.

3.3.2 Vertical Lift Performance Correlations

Figure 11.9 shows a vertical lift performance (VLP) correlation plot. For this work, Dons and Ros modified the VLP and Petroleum Experts' correlations were applied. The purpose of performing this plot is to ascertain which correlation best fits the wells under investigation. These results were obtained by plotting a test point on a depth-pressure transverse curve as indicated as the blue point at the bottom of Fig. 11.9.

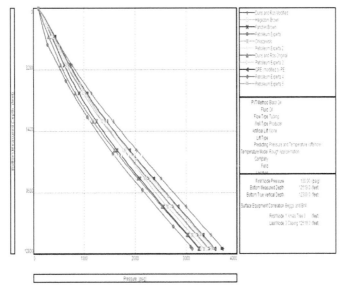

Fig. 11.9. Shows a vertical lift performance correlation plot.

3.3.3 Desired Flow Rates

The desired rate is the rate to achieve optimum production based on technical and economic factors. This is usually a choice of management. In this work, technical consideration was the basis to determine the desired flow rate. Desired flow rate is required for gas lift. A sensitivity plot was obtained by varying some parameters such as reservoir pressure, water cut and total gas-oil ratio for each well. Figure 11.10 shows the sensitivity plot for Well A displaying the test rate of 14,008 BOPD and pressure at 3,071 psig. The desired flow

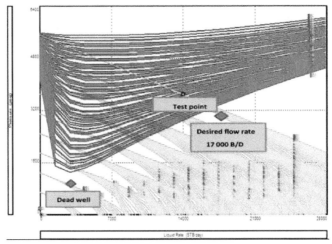

Fig. 11.10. TPR vs IPR Well A.

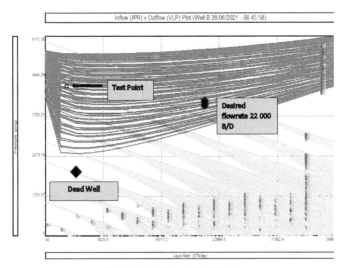

Fig. 11.11. TPR vs IPR Well B.

rate which is 17,000 BOPD (Fig. 11.10) still falls outside the intersection indicating that the rate of production for that well must be changed since that value is not optimum. It can be observed that the test point falls within where the outflow performance curve and inflow performance curve intersect meaning the desired test rate is feasible. It can be deduced that the well will cease to flow and become a dead well if the bottom hole pressure falls below 1,400 psi at the prevailing reservoir conditions since the inflow performance curves and the tubing performance curves cease to intersect. This observation was made for Well A. The sensitivity plot obtained for Well B indicates a maximum objective flowrate of 22,000 BOPD and a test rate of 3,026 STB/D at a pressure of 4,872 psi. From the graph in Fig. 11.11, it can be deduced that below 2,472 psi, Well B becomes a dead well since the inflow performance curves and the tubing performance curves cease to intersect.

3.4 Gas Lift Results

The results were obtained for the various gas lift modelling for Wells A and B and their respective gas sensitivity plots. The sensitivity plots show the optimum light gas injection rate as well as the optimum production rate for each well. For the parameters imputed into the simulation program, below are the results obtained for each of the wells. Figures 11.12 and 11.13 show the gas lift design for Wells A and B which were designed with four valves using the spacing line procedure. Figures 11.14 and 11.15 show the gas sensitivity plot for Wells A and B for the maximum economic gas to be injected.

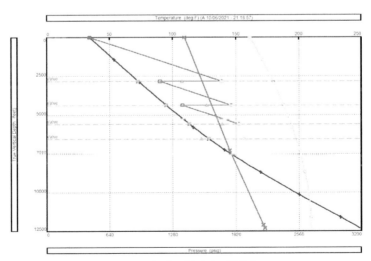

Fig. 11.12. Gas lift design Well A.

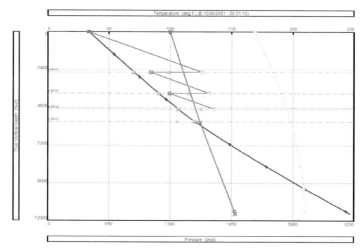

Fig. 11.13. Gas lift design Well B.

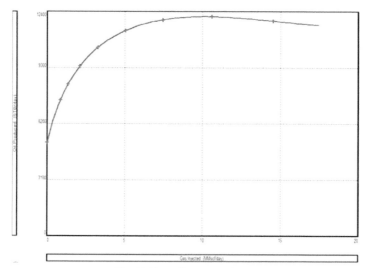

Fig. 11.14. Well A gas sensitivity.

3.4.1 Optimum Production Rates

Upon Simulation in PROSPER, the following production rates in Table 11.13 were obtained. Comparing the average daily base case flowrates calculated using the production history of the field to the results obtained below after applying gas lift modelling, there has been an increase in the deliverability of both wells. There was an 83.63% increase for Well A and a 61.64% increase for Well B.

Artificial Lift Design for Future Inflow and Outflow Performance for Jubilee Oilfield 279

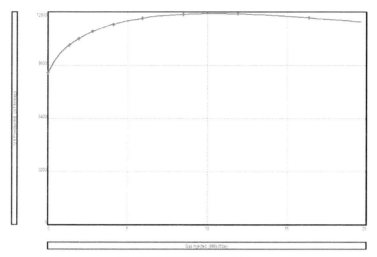

Fig. 11.15. Well B gas sensitivity.

Table 11.13. Gas lift results for Well A and B (Optimum Production Rates).

Well	Liquid Rate (STB/D)	Oil Rate (STB/D)	Injected Rate (MMSCF)	Injection Pressure (Psig)
A	17432.2	14991.7	5.47332	1400
B	19903.1	15922.5	6.97933	1500

3.5 ANN Results

3.5.1 ANN Architecture

Table 11.14 shows the statistics of the historical production data input and output for both wells used for the ANN models to predict THP. However, production data of three wells were used to build the database for training the model while one well data was used for testing the model.

Table 11.14. Input and output data statistics.

Parameters	Minimum	Maximum	Average	Standard Deviation	Data Type
Days	0	2455	1009.04	660.53	Input
Daily Oil Production (MMSTB/D)	0	27645	11308.43	5023.96	Input
Cumulative Oil Production (MMSTB)	0	26.06	10.44	6.68	Input
Cumulative Gas Production (MMSCF)	0	32.73	8.04	9.41	Input
Gas Rate (MMSCF/D)	0	23.66	5.94	6.51	Input
Flowing Bottomhole Pressure (PSIA)	2489.96	6212.41	4423.02	508.60	Input
Tubing Head Pressure (PSIG)	1095.03	4185.29	2564.75	430.48	Output

3.5.2 Model Visualisation

Figure 11.16 shows the training plot whereas Fig. 11.17 shows the testing plot obtained for the BPNN model. The results indicate that the prediction points are closer to the ideal line with an architecture of five inputs, one neuron, and one output. BPNN shows how best the prediction results are close to the actual. It can also be observed that the testing well dataset generalises very well on the training model.

Figure 11.18 shows the training plot whereas Fig. 11.19 shows the testing plot obtained for the RBFNN model. The testing RBFNN shows how best the prediction results are close to the actual with an architecture of six inputs, 17 neurons, and one output. The testing RBFNN shows how best the prediction results are close to the actual. Also, it can also be observed that the testing well dataset generalises very well on the training model.

RBFNN is a powerful AI technique used in well performance prediction, and it demonstrated better prediction accuracy than the BPNN model in this study. However, RBFNN has some limitations that could impact the accuracy of the results. One of the primary limitations of RBFNN is its sensitivity to the number and placement of basis functions, which are critical to the accuracy of the model. The selection of the appropriate number and location of basis functions can be challenging, and an improper choice can lead to overfitting or underfitting of the data. Additionally, the choice of the radial basis function itself could impact the accuracy of the model. The potential impact of these limitations on the results of the study is that the accuracy of the RBFNN model

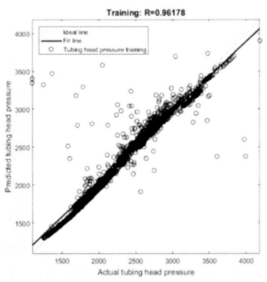

Fig. 11.16. Training plot for BPNN model.

Artificial Lift Design for Future Inflow and Outflow Performance for Jubilee Oilfield 281

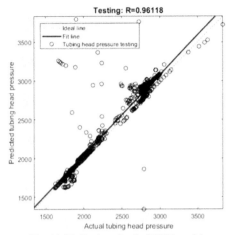

Fig. 11.17. Testing plot for BPNN model.

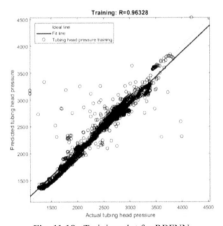

Fig. 11.18. Training plot for RBFNN.

could be influenced by the choice of basis functions and the quality of the data used in the model training. Therefore, it is essential to carefully evaluate the quality of the data used and the choice of basis functions to ensure the accuracy of the RBFNN model predictions. Additionally, further research can explore the use of other AI techniques to complement RBFNN and improve the accuracy of the predictions

Table 11.15 shows the results compilation for the ANN with the accuracy of each model and the computational time. RBFNN was the best model when tested with unseen data while BPNN obtained the fastest computational time due to the one neuron used than the 17 neurons by RBFNN.

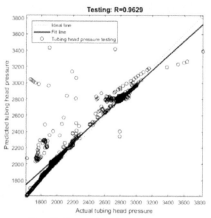

Fig. 11.19. Testing plot for RBFNN.

3.6 Discussion

The gas lift models developed in this study using PROSPER simulation software have significant implications for the accuracy of well performance predictions. The application of the gas lift model for both wells helps to optimise the performance of the artificial lift system, resulting in improved well productivity and efficiency. There was an 83.63% increase for Well A and a 61.64% increase for Well B. By using historical production data, the gas lift model can provide accurate predictions of future well performance under different operating scenarios. This is important for optimising production schedules, as well as for identifying potential production issues and troubleshooting them in advance. The gas lift models also enable the prediction of downhole and surface parameters, such as tubing head pressure and gas injection rate, which are critical in determining the effectiveness of the gas lift system. The development of an optimised gas lift model has significant implications for the oil and gas industry. It can lead to more efficient and cost-effective production, improved asset management, and increased profits. Furthermore, the methodology developed in this study can be applied to other oil fields with similar production characteristics, providing a standardised approach to optimising artificial lift systems. Overall, this research highlights the potential benefits of integrating AI techniques into gas lift modelling for accurate well performance predictions and optimisation. While the findings of this study are specific to the Jubilee Field, the methodology developed can be applied to other oil fields with similar production characteristics. The recommendation of implementing artificial lift at the start of production for nearby fields, such as the Pecan Field, is based on the observed technical problems faced in the Jubilee Field and the potential benefits of early

Table 11.15. Statistical measures used for ANN models' assessment.

MODELS (ANN)	BPNN	RBFNN
Training R-Correlational Coefficient	0.96178	0.96328
Mean Absolute Percentage Error (MAPE)	0.01651	0.02055
Testing R-Correlational Coefficient	0.96118	0.96290
MAPE	0.02764	0.02134
Computational Time (seconds)	6.83160	10.7950

intervention. However, the applicability of this recommendation to other fields should be evaluated based on their unique characteristics and conditions.

4. Conclusions

In summary, this research has contributed to the development of an optimised artificial lift system for the Jubilee Field, resulting in increased deliverability of all wells. Additionally, the application of AI techniques, such as BPNN and RBFNN, have demonstrated improved prediction accuracy for well performance. These findings have important implications for the optimisation of oil field production, and the methodology developed in this study can be applied to other fields to improve efficiency and productivity. The following conclusions can be drawn from the study.

- Successfully screen and design an optimum artificial lift system (gas lift), for the field under investigation.
- Comparing the average daily base flow rates calculated using the production history of the field to the results obtained after applying gas lift modelling with an increase in the deliverability of all the wells. There was an 83.63% increase for Well A and a 61.64% increase for Well B.
- RBFNN gave the best prediction for the tubing head pressure in the Jubilee Field than the BPNN model used with an R of 96.290% and MAPE of 0.02764.
- BPNN had less computational time due to the number of neurons used as compared to the RBFNN model.
- Artificial lift method is recommended at the start of production for the field nearby of Jubilee Field which is the Pecan Field operated by Aker Energy to avoid technical problems faced in the Jubilee Field.

Acknowledgment

We would like to thank anonymous reviewers for their contribution to this research. We are also grateful to Ghana National Petroleum Corporation for providing access to data for this research to be successful. We also acknowledge PETEX for making the educational licence of the software available. Finally, we thank the University of Mines and Technology, Tarkwa, Ghana for their immense support.

References

Agyeman, B.K. 2020. *Design of Gas Lift and Electric Submersible Pump: A Case Study at Jubilee Field*. Unpublished BSc Project Report, University of Mines and Technology, Tarkwa, pp. 1–118.

Akwensi, P.H., Brantson, E.T., Niipele, J.N. and Ziggah, Y.Y. 2021. Performance evaluation of artificial neural networks for natural terrain classification. *Applied Geomatics* 13(3): 453–465.

AL-Dogail, A.S., Baarimah, S.O. and Basfar, S.A. 2018. Prediction of inflow performance relationship of a gas field using artificial intelligence techniques. *In*: SPE Kingdom of Saudi Arabia Annual Technical Symposium and Exhibition. April. OnePetro.

Baudoin, C.R. 2016. Deploying the industrial Internet in oil & gas: Challenges and opportunities. *In*: SPEntelligent Energy International Conference and Exhibition. OnePetro.

Bhatia, A. and McAllister, S. 2014 *Artificial Lift: Focus on Hydraulic Submersible Pumps*, ClydeUnion Pumps, Tech 101(10): 3, 29–31.

Boyun, G., William, C.L. and Ali, G. 2007. *Petroleum Production Engineering, A Computer Assisted Approach*. Oxford, UK: Elsevier Ltd., pp. 186–300.

Brantson, E.T., Ju, B., Omisore, B.O., Wu, D., Aphu, E.S. and Liu, N. 2018. Development of machine learning predictive models for history matching tight gas carbonate reservoir production profiles. *Journal of Geophysics and Engineering* 15(5): 22–35.

Brantson, E.T., Ju, B., Opoku Appau, P., Akwensi, H.P., Agyare Peprah G., Liu, N., Aphu, E.S., Annan Boah, E. and Aidoo Borsah, A. 2019a. Development of low salinity water polymer flooding numerical reservoir simulator and smart proxy modeling for hybrid chemical enhanced oil recovery (CEOR). *Journal of Petroleum Science and Engineering* 187: 1–23.

Brantson, E.T., Ju, B., Ziggah, Y.Y., Akwensi, P.H., Sun, Y., Wu, D. and Addo, B.J. 2019b. Forecasting of horizontal gas well production decline in unconventional reservoirs using productivity, soft computing and swarm intelligence models. *Natural Resources Research* 28(3): 717–756.

Brantson, E.T., Osei, H., Aidoo, M.S.K., Appau, P.O., Issaka, F.N., Liu, N., Ejeh, C.J. and Kouamelan, K.S. 2022. Coconut oil and fermented palm wine biodiesel production for oil spill cleanup: Experimental, numerical, and hybrid metaheuristic modelling approaches. *Environmental Science and Pollution Research*, 1–19.

Brill, J.P. and Beggs, H.D. 1991. *Two-Phase Flow in Pipes*, Houston, 6th Edition, 1991, pp. 1–15.

Brown, K.E. 1982. Overview of artificial lift systems. *Journal of Petroleum Technology*, SPE 9979-PA, 13 pp.

Chen, X., Wang, D.Y., Tang, J.B., Ma, W.C. and Liu, Y. 2021. Geotechnical stability analysis considering strain softening using micro-polar continuum finite element method. *Journal of Central South University* 28(1): 297–310.

Flatern, R.V. 2015. Oilfield Review, Artificial Lift Systems, www.slb.com, Accessed November 2019.

Fleshman, R. and Lekic, H.O. 1999. Artificial lift for high production. *Oilfield Review Spring* 49–63.

Hill, T. and Remus, W. 1993. Neural network models for intelligent support of managerial decision making. *Hawaii University* 11(5): 449–459.

Lea, J.F. and Nickens, H.V. 1999. Selection of Artificial Lift. *SPE* 52157, Oklahoma City, Oklahoma 30 pp.

Mohammed, A.G.H. and Nasr, G.G. 2016. Gas lift optimisation to improve well performance., *World Academy of Science, Engineering and Technology, International Journal of Mechanical and Mechatronics Engineering* 10(3): 512–520.

Neely, B., Gipson, F., Capps, B., Clegg, J. and Wilson, P. 1981. Selection of Artificial Lift Method. *SPE* 10337 Dallas, Texas. 1 p.

Pennel, M., Hsiung, J. and Putcha, V.B. 2018. Detecting failures and optimising performance in artificial lift using machine learning models. *In: SPE Western Regional Meeting*. OnePetro.

Renpu, W. 2011. Selection and Determination of Tubing and Production Casing Sizes. *Advanced Well Completions Engineering*, 3rd Edn.. Houston,: Gulf Professional Publishing, pp. 117–170.

Schempf, F.J. 2011. Jubilee brings Ghana to the fore among West African deepwater regions. *Offshore (Conroe, Tex.)* 71(4).

Takacs, G. 2015. *Sucker-rod Pumping Handbook: Production Engineering Fundamentals and Long-stroke Rod Pumping*. Houston: Gulf Professional Publishing.

Woods, J.D. and Lea, J.F. 2017. What Is New In Artificial Lift? *World Oil Production Technology: Artificial Lift*, USA: Gulf Publishing, pp. 43–50.

Wu, D., Ju, B. and Brantson, E.T. 2016. Investigation of productivity decline in tight gas wells due to formation damage and Non-Darcy effect: Laboratory, mathematical modelling and application. *Journal of Natural Gas Science and Engineering* 34: 779–791.

CHAPTER 12

Modelling Two-phase Flow Parameters Utilizing Machine-learning Methodology

Longtong Dafyak* and Buddhika Hewakandamby

1. Introduction

Two-phase flow is predominant in natural occurring and industrial processes; from water droplets entrained in the air to domestic water distribution lines, combustion engines, power generation plants, amongst others. In the exploration industry, gas-liquid flow in pipes is a standard practice due to the coexistence of these fluids in the subsurface (McCain, 1994; Soloveichik et al., 2022). Furthermore, pipelines are still the safest and most economical means of transporting fluids over long distances (Green and Jackson, 2015; Canada Energy Report, 2020).

When gases and liquids flow simultaneously in pipes, unique flow configurations are formed, which are referred to as flow patterns. These flow patterns were classified as bubbly, dispersed bubbly, plug, slug, stratified, and annular flows by Hewitt and Roberts (1969) and Mandhane et al. (1974) as illustrated by Shoham (2006) in Fig. 12.1. The flow patterns formed depend on the operating parameters, pipe geometric variables, and physical properties of the fluids (Lu, 2015). Each flow pattern is associated with specific interaction between the phases and the pipe walls, thus making one flow pattern entirely different from another (Nie et al., 2022). The difference in momentum, phase distribution, and velocity distribution, makes gas-liquid flow dynamics a lot more complex than single-phase flow. However, some flow patterns share similar intermittent behaviour and characteristic parameters.

University of Nottingham, UK.
* Corresponding author: longtongdafyak@gmail.com

Modelling Two-phase Flow Parameters Utilizing Machine-learning Methodology 287

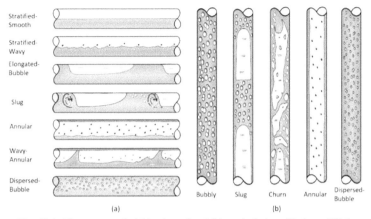

Fig. 12.1. Flow patterns in (a) horizontal and (b) vertical pipes (Shoham, 2006).

The wide application of gas-liquid flow and the curiosity of gas-liquid enthusiasts has inspired a plethora of studies in this field leading to the development of several gas-liquid models. Pioneering researchers (Nicklin, 1962; Dukler and Hubbard, 1975; Fernandes et al., 1983; Alves and Shoham, 1991) proposed physics-based models to describe two-phase flow dynamics; however, these models are either highly simplified or require high computation time to derive a solution. For these reasons, several empirical and semi-empirical models have been developed over the years to predict two-phase flow parameters. These correlations predict the mean void fraction and velocities for most flow patterns, the frequency, void fraction, and length of flow structures for intermittent flow patterns amongst other two-phase flow parameters. There are primarily two limitations of these empirical/semi-empirical correlations; one of which is the constraint of its applicability to flow conditions from which it is developed. Second, some of these correlations have proven to be applicable over a wide range of two-phase flow conditions; however, the simplistic curve fitting approach used in developing most empirical correlations are often inadequate for optimally modelling the complex relationships between the two-phase flow variables.

As Machine Learning (ML) application develops and gains popularity, researchers have explored ML techniques in predicting two-phase flow patterns and parameters. Flow pattern models developed by Cai et al. (1994), Rosa et al. (2010), Al-Naser et al. (2016), Mask et al. (2019), Liu et al. (2020), and dos Santos Ambrosio et al. (2022) using support vector machine, artificial neural networks (ANNs) and tree-based algorithms show promising results with an accuracy of approximately 96–100%. Yan et al. (2018) reviewed soft computing techniques, ML, evolutionary computation, fuzzy logic, and probabilistic reasoning, and proposed the trajectory of future development in multiphase flow research. Azizi et al. (2016) utilized ANN to predict liquid fraction using 468 experimental data points for oil-water mixtures. Kim et al. (2020) and Abdul-Majeed et al. (2022) proposed ML models to specifically model slug flow parameters. More recently, the ANN model proposed by Aliyu et al. (2023) for entrained liquid fraction in annular gas-liquid flows show good performance with a Root Mean Squared Error (RMSE) and R^2 of 0.005 and 0.97, respectively. These studies also compared existing correlations and demonstrated the potential of ANN, random forest, and support vector regression to outperform traditional two-phase flow correlations. Despite the contributions of these studies to data-driven modelling, studies focused on comparing the capabilities of varying ML algorithms in two-phase flow modelling is sparse. Thus, this study is focused on evaluating the performance of several ML algorithms in two-phase flow predictions and assessing the applicability of these ML models in comparison to traditional empirical correlations.

In this study, the interest is limited to the mean void fraction and structure velocity of intermittent flow regimes. This is because intermittent flows (slug, plug/elongated bubble, and churn) are the most persistent and often the extremely challenging flow patterns in two-phase flow systems. These two parameters are also generic characteristics of most two-phase flow patterns; thus, the developed models are applicable across a wide breadth of two-phase flow conditions. The two-phase flow variables considered in this study are the operating conditions (the gas and liquid superficial velocities, the pipe diameter and inclination) and the fluid properties (the gas and liquid viscosity, the gas and liquid density).

2. Data Sources and Existing Correlations

The data utilized in this study was acquired from experiments conducted by Escrig (2017) and Dafyak (2022) in the gas-liquid flow facility at University of Nottingham. The experimental procedure is elaborately described in the studies by Escrig et al. (2017) and Dafyak et al. (2021). Table 12.1 details the data sources and the corresponding operating variables for the data utilized in this study for mean void fraction and structure velocity.

Table 12.1. Data sources and summary of operating variables.

Author (year)	Number of Data Points	Fluid System	D	θ	U_{SL}	U_{SG}	ρ_L	μ_L
			(mm)	(°)	(m/s)	(m/s)	(kg/m³)	(cp)
Dafyak (2022)	985	Air-oil	67	0–90	0.03–2.82	0.08–0.3	928	93–230
Escrig (2017)	493	Air-oil	67	0–90	0.06–2.92	0.02–0.47	917	5

Both experiments were conducted in a 67 mm pipe for air and silicone oil mixtures with a similar range of pipe inclination, gas and liquid velocities; however, the fluid properties are different. A total of 1,478 experimental runs were conducted at varying pipe inclination, superficial gas and liquid velocities, and liquid viscosities. The data set covers upward inclined flows between the horizontal and vertical axis (0°–90°) for low to medium liquid viscosities; Escrig (2017) explored 7 pipe inclinations whereas Dafyak (2022) covered 10 pipe inclinations within this range.

Table 12.2 shows a collation of some of the most widely used correlations for predicting the mean void fraction and structure velocity. These empirical correlations developed on the basics of slip-ratio and drift-flux model are semi-empirical and have proven to be robust and easily scalable, showing good performance over a wide range of fluid and flow parameters (Woldesemayat and Ghajar, 2007; Godbole et al., 2011). For this reason, the existing correlations considered in this study are limited to slip-ratio and drift-flux correlations for both the mean void fraction and the structure velocity. The performance of these correlations is evaluated using the test data for this study in Section 5.

3. Methodology

The sequential procedure for data processing and model development in this study are data acquisition and sourcing, data pre-processing, model development, hyperparameter tuning and model evaluation, and validation. All analysis was carried out in the Python environment using open-source libraries; Pandas, Numpy, Matplotlib, and Seaborn for data manipulation and visualization and Keras and Sklearn for predictive modelling.

The experimental data for this study was sourced from previous studies as elaborated in Section 2. Data pre-processing and feature engineering encompasses all the steps taken to prepare the data for optimal model development and evaluation. The input parameters referred to as features and output parameters referred to as target values in this study were identified (Fig. 12.2). Correlation analysis is utilized to identify the magnitude of the relationship between input variables and target variables. This helps in selecting the most relevant variables for model development, eliminate irrelevant variables to reduce noise, avoid multicollinearity, and improve the overall performance of the model. Prior to model development, the data was split into training and test data for model development and model evaluation, respectively. The same number of data were used to train and test all the ML algorithms considered in this study. The supervised learning algorithms utilized in this study are Multiple Linear Regression (MLR), Polynomial Regression (PR), Random Forest (RF), Support Vector Regression (SVR),

Table 12.2. Existing empirical correlations.

Authors	Correlations
Mean void fraction	
Lockhart, R.W., Martinelli (1949)	$\varepsilon = \left[1 + 0.28\left(\dfrac{1-x}{x}\right)^{0.64}\left(\dfrac{\rho_G}{\rho_L}\right)^{0.36}\left(\dfrac{\mu_L}{\mu_G}\right)^{0.07}\right]^{-1}$
Bendiksen (1984)	$\varepsilon = \dfrac{U_{SG}}{1.2U_m + 0.35\sin\theta\sqrt{gd} + 0.54\cos\theta\sqrt{gd}}$
Rouhani and Axelsson (1970)	$\varepsilon = {x}/{\rho_G}\left[C_o\left(\dfrac{x}{\rho_G} + \dfrac{1-x}{\rho_L}\right) + \dfrac{U_{GM}}{G}\right]^{-1}$ Where $U_{GM} = \left(\dfrac{1.18}{\sqrt{\rho_L}}\right)(g\sigma(\rho_L - \rho_G))^{0.25}$ for P < 12.7
Premoli et al. (1970)	$\varepsilon = \left[1 + A_{PRM}\left(\dfrac{1-x}{x}\right)^a\left(\dfrac{\rho_G}{\rho_L}\right)^b\right]^{-1}$ Where $A_{PRM} = 1 + F_1\left[\dfrac{y}{1+yF_2} - yF_2\right]$, $F_1 = 1.578Re_L^{-0.19}\left(\dfrac{\rho_G}{\rho_L}\right)^{0.22}$, $F_2 = 0.0273We_L Re_L^{-0.51}\left(\dfrac{\rho_G}{\rho_L}\right)^{-0.08}$ $y = \left[\left(\dfrac{1-x}{x}\right)\left(\dfrac{\rho_G}{\rho_L}\right)\right]^{-1}$, $We_L = \dfrac{G^2 D}{\sigma\rho L}$, $Re_L = \dfrac{GD}{\mu L}$
Woldesemayat and Ghajar (2007)	$\varepsilon = \dfrac{U_{SG}}{B + C}$ $B = U_{SG}\left(1 + \left(\dfrac{U_{SL}}{U_{SG}}\right)^{\left(\frac{\rho_G}{\rho_L}\right)^{0.1}}\right)$ $C = 2.9\left[\dfrac{gD\sigma(1+\cos\theta)(\rho_L - \rho_G)}{\rho_L^2}\right]^{0.25}(1.22 + 1.22\sin\theta)^{\frac{P_{atm}}{P_{system}}}$
Structure velocity	
Bendiksen (1984)	$U_{TB} = 1.2U_m + 0.35\sin\theta\sqrt{gd} + 0.54\cos\theta\sqrt{gd}$
Hasan and Kabir (1988)	$U_{TB} = C_o U_M + U_{D(Vertical)}\sqrt{\sin\theta}(1+\cos\theta)^{1.2}$
Woldesemayat and Ghajar (2007)	$U_{TB} = B + C$

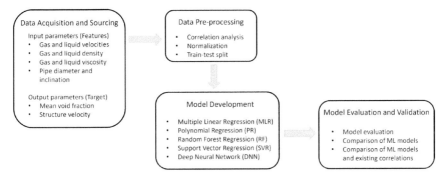

Fig. 12.2. Workflow for data processing and model development.

and Deep Neural Network (DNN). The characteristics of these algorithms are summarized in Table 12.3. The hyperparameters for each of the ML algorithms were tuned to optimize model performance and improve accuracy (Table 12.4). Selecting the optimal set of modelling parameters minimizes overfitting, optimizes training speed, and overall model generalization.

In the final step, the performance of each ML model was evaluated using the following statistical parameters: RMSE, Absolute Average Percentage Error (AAPE), and the Average Percentage Error (APE), and correlation coefficient (R^2). The performance of the ML models and existing correlations are also compared to evaluate the applicability of ML in modelling the two-phase flow parameters.

4. Results and Discussions

4.1 Data Pre-processing

Figure 12.3 shows the Pearson standard correlation coefficients between the flow variables and two-phase flow parameters. The pipe diameter, gas density, and viscosity showed the least correlation; this can be linked to the constant values of these parameters as air is used as the gas phase across all experimental points and all experiments were conducted in the same pipe (diameter = 0.067 m). Some existing models show a strong correlation between mixture velocity and the two-phase flow parameters considered in this study.

For this reason, the mixture velocity (superficial gas + superficial liquid velocities) is included as one of the variables in this study. The negative correlations indicate (U_{SL}, ρ_L, θ, and μ_L) an inverse relationship, whereas U_{SG} and U_m show a positive relationship with the mean void fraction. The U_{SG} shows the highest correlation for both the mean void fraction and the structure velocity. Although the correlation matrix in Fig. 12.3 shows a minimal correlation between the pipe inclination and the mean void fraction,

Table 12.3. Brief description of regression algorithms.

ML Algorithm	Illustration
Linear regression assumes a linear relationship between a dependent variable and independent variable(s). Linear regression can be either simple (one dependent variable) or multiple (2 or more independent variables). Linear regression estimates the coefficients (slopes and intercept) that minimize the loss function between the actual values and predicted values (Montgomery et al., 2012).	
Polynomial regression models nonlinear relationships between dependent and independent variables. Polynomial models can fit complex relationships by adding higher-order terms; however, the appropriate degree of polynomial is an important consideration to minimize overfitting (James et al., 2013).	
Random forest is an ensemble learning method developed as a combination of multiple decision trees. RF models are developed following this sequence: Bootstrapping and tree building – randomly sampling of the original dataset to create multiple subsets of the data. Pruning – removing nodes or branches that do not improve the model's performance. Majority voting – aggregates the predictions of multiple decision trees to make a final prediction (Breiman, 2001).	
Support vector regression learns the function that maps input features to continuous output values by finding a hyperplane that maximally separates the input data while minimizing the error on the training data. SVR uses kernel functions to transform input data into a higher dimensional space, where it is easier to find a linear decision boundary (Chang and Lin, 2012).	
The architecture of a **deep neural network** consists of multiple layers of nodes. Each node performs a different type of computation and learns complex relationships between input and output variables. The first layer of the network receives the input variables, and subsequent layers process the data while the output layer provides the final prediction. The weight and bias of the nodes in the network are iteratively adjusted to minimize the difference between the predicted and the actual output (Goodfellow et al., 2016).	

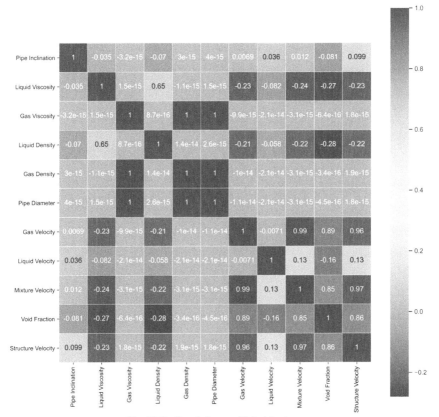

Fig. 12.3. Correlation coefficient heatmap.

the empirical correlations by Bendiksen (1984) and Woldesemayat and Ghajar (2007) suggest otherwise; thus, the pipe inclination is included in the modelling data set for this study. The air properties and pipe diameter were eliminated from the dataset as both variables show no correlation with the output data. The pipe inclination, superficial gas and liquid velocities, liquid viscosity and density showed some effect on the target variables; thus, all these flow variables were utilized in the model development to capture their impact on the mean void fraction and structure velocity. The entire data was split in a train-test ratio of 4:1; that is, 1,146 train data and 287 test data. It is worth noting that prior to model development, z-scores normalization was utilized to transform all the variables of the data to a similar scale.

4.2 Model Development and Evaluation

For each of the models, the training parameters were tuned to ensure optimal model performance while minimizing overfitting. For instance, several

Table 12.4. Training hyperparameters.

ML Algorithms	Parameters	Values
PLR	Polynomial degree	4
RF	Number of decision trees	100
	Depth of decision tree	20
SVR	Kernel	rbf
DNN	Input layer activation function	relu
	Hidden layers activation function	tanh
	Kernal initializer for all layers	normal
	Optimizer	adam
	Batch size	20
	Number of epochs	50

RF models were developed by varying the maximum number of trees and maximum depth within the range of 50–300 and 20–50, respectively. The model with the optimal performance and enough trees to ensure randomness in the bootstrapping and tree-building stages of the RF model development. The set maximum number of trees and tree depth in this case are 100 and 20 as shown in Table 12.4. Similar hyperparameter tuning techniques were replicated for PR, SVR, and DNN and the optimal values are presented in Table 12.4.

Table 12.5 shows the mean void fraction and structure velocity model performances for the test data based on the predefined statistical error parameters for each of the ML models. All the ML models, except multiple linear regression, show an acceptable prediction for mean void fraction. Although the performance of the nonlinear models for the mean void fraction in all is above 90%, RF and DNN show superior predictive capabilities in comparison to PR and SVR. As shown in Table 12.5, DNN shows the best R^2 and RMSE while RF has the best APE and APPE; however, these values are within similar range for both ML algorithms. For the structure velocity, all the nonlinear models show similar capability with an RMSE of 0.1, APPE and R^2 ranging between 3.5–3.7% and 99.0–99.2%, respectively. The MLR model for U_{TB} shows acceptable performance with an RMSE and R^2 of 0.24 and 94.7; however, it is not comparable to the other ML models considered in this study. For this reason, the MLR models for both the mean void fraction and structure velocity are considered unsuitable for further analysis due to their poor performance in comparison to the other ML models. Figure 12.4 shows the cross plots of the predicted parameters against the test data for each of the ML models. For the void fraction, the predicted values all align with 20% of the test data for RF and DNN. Figure 12.4, eludes the poor performance of MLR

Table 12.5. Statistical error parameters for ML models.

ML Algorithms	RMSE (-)	APE (%)	APPE (%)	R^2 (%)
Mean void fraction				
Multilinear regression	0.08	24.92	9.49	83.81
Polynomial regression	0.02	3.98	−0.29	98.37
Random forest	0.02	2.21	0.27	99.32
Support vector regression	0.02	6.30	−0.61	98.60
Deep neural network	0.01	2.86	−0.60	99.55
Structure velocity				
Multilinear regression	0.24	1.84	9.77	94.74
Polynomial regression	0.10	0.36	3.69	98.96
Random forest	0.10	0.61	3.52	99.01
Support vector regression	0.10	0.39	3.94	98.97
Deep neural network	0.10	−1.27	3.63	98.99

in prediction of the mean void fraction, sharing significant scatter especially for $\varepsilon > 0.4$. For the structure velocity, majority of the data falls within 20% for all the ML models. The outcome of the MLR model supports the good performance of linear correlations like Nicklin (1962) and Bendiksen (1984). All the models developed in this study were training in less than 2 minutes; thus, the training time was not considered as a performance criterion. However, the ANN was more computationally challenging as it required tuning several hyperparameters. For this reason, RF is selected as the best model for the data set used in this study even though it shows similar statistical error parameters with DNN.

5. Comparison between ML Algorithms and Existing Correlations

This subsection evaluates the performance of existing correlations and compares it to that of the ML algorithms evaluated in Section 4. The same test data is used for evaluating the ML models and the existing correlations. Table 12.6 shows that the Bendiksen (1984) and Woldesemayat and Ghajar (2007) models perform better than the other correlations with RMSE and R^2 of 0.04 and ~ 96%, respectively. These models cover all upward inclined flows. It is worth noting that the Lockhart and Martinelli (1949), Rouhani and Axelsson (1970), and Premoli et al. (1970) correlations were developed specifically for vertical flows only. Although Bendiksen's model does not consider the fluid properties, it presents similar capabilities to the Woldesemayat and Ghajar

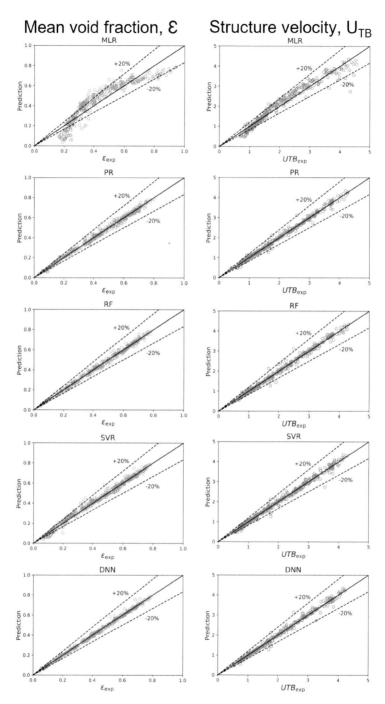

Fig. 12.4. Cross plot of predicted and test data for ε and U_{TB}.

Table 12.6. Statistical error parameters for empirical correlations.

Empirical Correlations	RMSE	APE	APPE	R^2
	(-)	(%)	(%)	(%)
Mean void fraction				
Lockhart and Martinelli (1949)	0.09	8.82	20.90	80.69
Bendiksen (1984)	0.04	−1.28	12.82	95.50
Rouhani and Axelsson (1970)	0.10	24.03	24.97	73.09
Premoli et al. (1970)	0.15	−0.81	29.35	46.97
Woldesemayat and Ghajar (2007)	0.04	7.54	8.77	96.24
Structure velocity				
Bendiksen (1984)	0.31	5.98	13.17	91.22
Hassan and Kabir (1988)	0.37	−14.99	17.18	87.41
Woldesemayat and Ghajar (2007)	0.37	−16.40	18.00	87.46

(2007) correlation which takes into account the surface tension, gas, and liquid densities.

Table 12.5 also shows the performance for the structure velocity correlations. Across all statistical error parameters, the Bendiksen (1984) correlation displays the best performance. For horizontal flows, drift flux correlations developed based on the assumption of zero drift velocity tend to predict the structure velocity with lower accuracy compared to correlations that assume the opposite (Woldesemayat and Ghajar, 2007). Thus, it is not surprising that the Bendiksen (1984) correlation performs better than the Hasan and Kabir (1988) correlation. Although the Woldesemayat and Ghajar (2007) correlation is developed based on the drift-flux model, assumes a non-zero drift velocity in the horizontal pipe, and considers some fluid properties, it fails to meet the predictive capabilities of the Bendiksen (1984) correlation for the data set explored in this study.

To further validate the ML models developed, only the best performing correlations are compared with the ML models development. For the experimental conditions and two-phase flow parameters explored in this study, the Bendiksen (1984) and Woldesemayat and Ghajar (2007) models show the best performance for both the structure velocity and mean void fraction. Comparing the statistical error parameters in Tables 12.4 and 12.5, all the ML algorithms, except MLR, outperform the traditional empirical correlations for predicting the mean void fraction. In the case of the structure velocity, all the ML models surpass the correlations considered in this study by at least 8% of the RMSE, APPE, APE, or R^2. This comparative assessment demonstrates the capabilities of ML algorithms in the development of data-driven models with substantially higher accuracy compared to the existing empirical models.

6. Conclusions and Recommendations

This study explores the capabilities of regression algorithms to predict two-phase flow parameters. The mean void fractions and structure velocities of intermittent flow patterns at varying gas and liquid superficial velocities, liquid viscosities and densities, and pipe inclination were predicted using multiple linear regression, polynomial regression, random forest, support vector regression, and deep neural network. For both parameters, the MLR predicted the target variables with the least accuracy, thus suggesting that the relationship between the two-phase parameters and flow variables are complex and cannot be adequately modelled using a linear relationship. Although all other ML models show a similar performance, RF and DNN show the best results with reference to the statistical error parameters considered in this study. A cross plot of the predicted results shows that majority of the data aligns within 20% of the corresponding test data. The ML predictive models were validated by comparing the performance with well-established correlations. All the ML models, except MLR, outperform the traditional empirical correlations considered in this study by at least 8% of the RMSE, APPE, APE, or R^2.

In this study, the ML models were evaluated using test data having the same configurations as the training data set. The authors suggest that future work should explore validation data from other sources to adequately validate the ML models, account for overfitting tendencies, and enable optimal hyperparameters tuning to optimize the versatility and generalization of these regression models. Furthermore, utilizing ML libraries to evaluate feature importance should be considered. This technique quantifies the contribution of each flow variable on the ML model and by extension, the magnitude of its influence on the two-phase flow parameter which is challenging to capture using the traditional correlation approach.

Nomenclature

A	area of pipe	x	quality
C_d	drift coefficient	θ	pipe inclination
C_o	distribution coefficient	ε	void fraction
D	diameter of the pipe	σ	Surface tension
g	acceleration due to gravity	ρ_L	liquid density
Re	Reynolds number	ρ_G	gas density
U_d	drift velocity	μ_G	gas viscosity
U_m	mixture velocity	μ_L	liquid viscosity
U_{TB}	translational velocity		

References

Abdul-Majeed, G.H., Kadhim, F.S., Almahdawi, F.H.M., Al-Dunainawi, Y., Arabi, A. and Al-Azzawi, W.K. 2022. Application of artificial neural network to predict slug liquid holdup. *International Journal of Multiphase Flow* 150(January): 104004. https://doi.org/10.1016/j.ijmultiphaseflow.2022.104004.

Al-Naser, M., Elshafei, M. and Al-Sarkhi, A. 2016. Artificial neural network application for multiphase flow patterns detection: A new approach. *Journal of Petroleum Science and Engineering* 145: 548–564. https://doi.org/10.1016/j.petrol.2016.06.029.

Aliyu, A.M., Choudhury, R., Sohani, B., Atanbori, J., Ribeiro, J.X.F., Ahmed, S.K.B. and Mishra, R. 2023. An artificial neural network model for the prediction of entrained droplet fraction in annular gas-liquid two-phase flow in vertical pipes. *International Journal of Multiphase Flow* 164: 104452. https://doi.org/10.1016/j.ijmultiphaseflow.2023.104452.

Alves, I.N. and Shoham, O. 1991. Slug flow phenomena in inclined pipes. *Petroleum Engineering*, University of Tulsa, USA. *Doctor of* (July 2007).

Azizi, S., Awad, M.M. and Ahmadloo, E. 2016. Prediction of water holdup in vertical and inclined oil-water two-phase flow using artificial neural network. *International Journal of Multiphase Flow* 80: 181–187. https://doi.org/10.1016/j.ijmultiphaseflow.2015.12.010.

Bendiksen, K.H. 1984. An experimental investigation of the motion of long bubbles in inclined tubes. *International Journal of Multiphase Flow* 10(4): 467–483. https://doi.org/10.1016/0301-9322(84)90057-0.

Breiman, L. 2001. Random forests. *Machine Learning* 45: 5–32.

Cai, S., Toral, H., Qiu, J. and Archer, J.S. 1994. Neural network based objective flow regime identification in air-water two phase flow. *The Canadian Journal of Chemical Engineering* 72(3): 440–445. https://doi.org/10.1002/cjce.5450720308.

Canada. Energy Report. 2020. *Oil Sands: Pipeline Safety*. https://www.nrcan.gc.ca/energy/publications/18754.

Chang, C.C. and Lin, C.J. 2012. LIBSVM: A library for support vector machines. *ACM Transactions on Intelligent Systems and Technology (TIST)* 2(3): 27.

Dafyak, L. 2022. *Hydrodynamics of Viscous Slug Flow in Inclined Pipes* (Issue December). University of Nottingham.

Dafyak, L.A., Hewakandamby, B., Fayyaz, A. and Hann, D. 2021. *Taylor Bubbles of Viscous Slug Flow in Inclined Pipes*. Paper presented at the Offshore Technology Conference, Virtual and Houston, Texas, August 2021. https://doi.org/10.4043/31238-MS.

dos Santos Ambrosio, J., Lazzaretti, A.E., Pipa, D.R. and da Silva, M.J. 2022. Two-phase flow pattern classification based on void fraction time series and machine learning. *Flow Measurement and Instrumentation* 83(November 2021): 102084. https://doi.org/10.1016/j.flowmeasinst.2021.102084.

Dukler, A.E. and Hubbard, M.G. 1975. A model for gas-liquid slug flow in horizontal and near horizontal tubes. *Industrial & Engineering Chemistry Fundamentals* 14(4): 337–347. https://doi.org/10.1021/i160056a011.

Escrig, J. 2017. *Influence of Geometrical Parameters on Gas-liquid Intermittent Flows*. PhD thesis, University of Nottingham.

Escrig, J., Hewakandamby, B. and Azzopardi, B. 2017. *Influence of Diameter and Inclination of the Pipes on the Velocity of Periodic Structures in Gas-Liquid Intermittent Flows*. November.

Fernandes, R.C., Semiat, R. and Dukler, A.E. 1983. Hydrodynamic model for gas-liquid slug flow in vertical tubes. *AIChE Journal* 29(6): 981–989. https://doi.org/10.1002/aic.690290617.

Godbole, P.V., Tang, C.C. and Ghajar, A.J. 2011. Comparison of void fraction correlations for different flow patterns in upward vertical two-phase flow. *Heat Transfer Engineering* 32(10): 843–860. https://doi.org/10.1080/01457632.2011.548285.

Goodfellow, I., Bengio, Y. and Courville, A. 2016. *Deep Learning* (Vol. 1). MIT Press.

Green, KP. and Jackson, T. 2015. Safety in the Transportation of Oil and Gas: Pipelines or Rail? *Fraser Institute*: 1–14.

Hasan, A.R. and Kabir, C.S. 1988. Predicting multiphase flow behavior in a deviated well. *SPE Production Engineering* 3(4): 474–482. https://doi.org/10.2118/15449-PA.

Hewitt, G.F. and Roberts, B.N. 1969. Studies of two-phase flow patterns by simultaneous X-ray and flash photography. *United Kingdom Atomic Energy Authority Research Group*, Berkshire.

James, G., Witten, D., Hastie, T. and Tibshirani, R. 2013. Introduction to Statistical Learning. *An Introduction to Statistical Learning with Applications in R*: 176–178.

Kim, T.W., Kim, S. and Lim, J.T. 2020. Modeling and prediction of slug characteristics utilizing data-driven machine-learning methodology. *Journal of Petroleum Science and Engineering* 195(June): 107712. https://doi.org/10.1016/j.petrol.2020.107712.

Liu, Y., Tong, T.A., Ozbayoglu, E., Yu, M. and Upchurch, E. 2020. An improved drift-flux correlation for gas-liquid two-phase flow in horizontal and vertical upward inclined wells. *Journal of Petroleum Science and Engineering* 195(June): 107881. https://doi.org/10.1016/j.petrol.2020.107881.

Lockhart, R.W. and Martinelli, R.C. 1949. *Proposed Correlation of Data for Isothermal Two-phase, Two-component Flow in Pipes* (pp. 45, 39–48). Chem. Eng. Progr. Univ. of Berkeley California.

Lu, M. 2015. *Experimental and Computational Study of Two-phase Slug Flow*. PhD Thesis, Imperial College of London, June.

Mandhane, J.M., Gregory, G.A. and Aziz, K. 1974. A flow pattern map for gas-iquid flow in horizontal pipes. *International Journal of Multiphase Flow* 1(4): 537–553. https://doi.org/https://doi.org/10.1016/0301-9322(74)90006-8.

Mask, G., Wu, X. and Ling, K. 2019. An improved model for gas-liquid flow pattern prediction based on machine learning. *Journal of Petroleum Science and Engineering* 183(August): 106370. https://doi.org/10.1016/j.petrol.2019.106370.

McCain, Jr. W.D 1933. *The Properties of Petroleum Fluids*. PennWell Publishing Company, Tulsa Oklahoma. ISBN 0-87814-335-1.

Montgomery, D.C., Peck, E.A. and Vining, G.G. 2012. *Introduction to Linear Regression Analysis*. Wiley.

Nicklin, D.J. 1962. Two-phase bubble flow. *Chemical Engineering Science* 17(9): 693–702. https://doi.org/https://doi.org/10.1016/0009-2509(62)85027-1.

Nie, F., Wang, H., Song, Q., Zhao, Y., Shen, J. and Gong, M. 2022. Image identification for two-phase flow patterns based on CNN algorithms. *International Journal of Multiphase Flow* 152(March): 104067. https://doi.org/10.1016/j.ijmultiphaseflow.2022.104067.

Premoli, A., Francesco, D. and Prima, A. 1970. An empirical correlation for evaluating two-phase mixture density under adiabatic conditions. In: *European Two-Phase Flow Group Meeting*, Milan, Italy.

Rosa, E.S., Salgado, R.M., Ohishi, T. and Mastelari, N. 2010. Performance comparison of artificial neural networks and expert systems applied to flow pattern identification in vertical ascendant gas-liquid flows. *International Journal of Multiphase Flow* 36(9): 738–754. https://doi.org/10.1016/j.ijmultiphaseflow.2010.05.001.

Rouhani, S.Z. and Axelsson, E. 1970. Calculation of void volume fraction in the subcooled and quality boiling regions. *International Journal of Heat and Mass Transfer* 13(2): 383–393. https://doi.org/10.1016/0017-9310(70)90114-6.

Shoham, Ovadia. 2006. *Mechanistic Modeling of Gas-liquid Two-phase Flow in Pipes*. Society of Petroleum Engineers. https://doi.org/10.2118/9781555631079.

Soloveichik, Y.G., Persova, M.G., Grif, A.M., Ovchinnikova, A.S., Patrushev, I.I., Vagin, D.V. and Kiselev, D.S. 2022. A method of FE modeling multiphase compressible flow in hydrocarbon reservoirs. *Computer Methods in Applied Mechanics and Engineering* 390: 114468. https://doi.org/10.1016/j.cma.2021.114468.

Woldesemayat, M.A. and Ghajar, A.J. 2007. Comparison of void fraction correlations for different flow patterns in horizontal and upward inclined pipes. *International Journal of Multiphase Flow* 33(4): 347–370. https://doi.org/10.1016/j.ijmultiphaseflow.2006.09.004.

Yan, Y., Wang, L., Wang, T., Wang, X., Hu, Y. and Duan, Q. 2018. Application of soft computing techniques to multiphase flow measurement: A review. *Flow Measurement and Instrumentation* 60(February): 30–43. https://doi.org/10.1016/j.flowmeasinst.2018.02.017.

Index

A

Akaike Information Criterion (AIC) 75
Analytical models 88
Anthropogenic CO_2 126, 127
Artificial Intelligence (AI) 1, 7, 35, 60, 106, 169, 182, 185, 208, 214, 261
Artificial lift methods 262, 272, 283
Artificial lift system 262, 263, 282, 283
Artificial Neural Networks (ANNs) 8, 208, 215, 220, 224, 261, 262, 267, 287

B

Backpropagation network 261
Backpropagation neural network 227, 262
Bayes theorem 75
Bayesian optimisation 1, 2
Bound fluid 33, 36–38, 40, 41
Bound fluid volume 36, 37
Brownfields 125

C

Capital and operating costs 105
Carbon dioxide 3
Carbon Dioxide (CO_2) injection 126
Climate change 126, 127
CO_2 flooding 127, 128
CO_2 storage 126, 127, 163
Compressional sonic 58, 61–63, 82
Computer vision 1

D

Data acquisition 105, 107, 110
Data analysis 11, 110, 163
Data mining framework 88
Data science 1, 4, 5
Data-driven models 88, 94–97, 100

Deep Adaptation Neural Network (DaNN) 161, 162
Deep Convolutional Neural Network (DCNN) 3, 159, 164–166, 168, 169, 172, 173, 179, 181, 182, 189
Deep learning 39, 43, 160, 161, 163, 176, 178, 179
Deep learning-based algorithms 160
Deep neural network 250, 252
Distributed temperature sensing systems 105
Downhole control systems 105
Downhole sensors 104
Drilling 6–11, 17, 20, 23, 26–29
Drilling fluid 6–9, 11, 27, 28
Drilling operations 7, 10, 11

E

Electrical submersible pump 88, 90, 92, 100
Empirical correlation 58, 64, 75, 79–82, 105, 106, 111, 208, 248, 287–290, 293, 297, 298
Enhanced Oil Recovery (EOR) 3, 125
Ensemble models 33, 43–45, 48, 49, 54
Evolutionary computation 287
Explainable AI 9, 15, 28, 29, 110, 114

F

Facies analysis 208, 234, 235, 250
Faults 159, 160, 162, 169
Flow dynamics 286, 287
Flow metering 87, 88, 96
Flow pattern models 287
Flow rate 87, 88, 90, 92–97, 99, 100
Flow structures 287
Fluid distribution 105
Fluid phase properties 87
Fluid sequestration 57, 58, 83
Forecasting 261, 262

Formation evaluation 57
Fossil fuel combustion 127
Fractal analysis 213
Free fluid 33–35, 37, 38, 40, 41, 47, 48, 54
Free fluid index 33
Free fluid volume 33–35, 47, 48, 54
Fuzzy logic 261, 287

G

Gas injection 125, 127, 129, 141, 142, 145
Gas lift 262, 264, 265, 267, 268, 270–273, 276–279, 282, 283
Gas oil ratio (GOR) 87
Gas reservoir 261
Gas-liquid flow 286–288
Geochemical modelling 132, 153
Geological formations 159
Geomechanical properties 57
Geophysics 57
Geothermal energy 58
Ghana 261, 262, 269, 284
Global energy demand 125
Global optimisation 75
Global warming 127
Greenhouse gas emissions 126, 127
Greenhouse gases 126
Group method of data handling 132, 136

H

Heterogeneous formation 104
Hydrocarbon extraction 33
Hyperparameter tuning 75

I

Immiscible CO_2 flooding 127
Inflow and outflow performance 261
Inflow Control Valves (ICVs) 105, 107, 108, 122
Inflow performance relationship 263
Intermittent flow 287, 288, 298
Interval control devices 105

J

Jubilee field 262, 268–270, 282, 283

L

Liquid rates 87
Lost circulation 6–9, 11, 26–28
Low Salinity Water (LSW) 126

M

Machine Learning (ML) 1–5, 6–11, 13, 15, 16, 18, 23, 26, 36, 38, 43, 47, 57, 58, 60, 61, 64, 68, 73, 104, 106, 108–111, 113, 114, 122, 160, 161, 178, 287–289, 291, 292, 294, 295, 297, 298
Machine learning algorithms 97
Maximum likelihood 75
Mean void fraction 287–291, 293–295, 297, 298
Mercury Injection Capillary Pressure (MICP) analysis 34
Metaheuristics algorithms 64
Miscible CO_2 flooding 127, 128
Model agnostic 6, 9, 15, 16, 20, 22–24, 28
Mud loss 7, 9, 10, 12, 13, 18–20, 22–25, 27, 28
Multioutput supervised machine learning 58
Multiphase flowrates 87
Multiphase physical flow meters (MPFMs) 87
Multivariate adaptive regression splines 132, 134
Multi-zonal reservoirs 105, 106

N

NMR porosity 36, 39
Nonconventional wells 104
Nuclear Magnetic Resonance (NMR) 34, 35

O

Objective functions 65, 66, 75
Oil and gas industry 160, 161
Oil and gas production systems 87
Oil and gas reservoir production 105
Oil recovery 261
Oil recovery factor 125, 131–134, 136–138, 142, 153, 154
Oil reserves 125
Oil swelling 126
Original Oil in Place (OOIP) 126

P

Permanent monitoring 105, 107
Permeability prediction 209, 223, 230, 231, 235, 240, 247
Petroleum reservoir 125
Phase fractions 87, 88
Physics-based models 287
Pore fluids 33, 34, 37, 38, 57
Pore pressure 57

Pore volume fraction 33, 35, 36, 54
Porosity 33–42, 47–50, 52–54
Porosity prediction 230
Predictive models 93–95
Pressure 104–108, 110, 111, 115, 116
Pressure control valves 104
Primary recovery 125, 126, 130
Probabilistic model 75
Probabilistic reasoning 287
Production rate 261–263, 277–279
Production systems 87–91, 99
Production testing 87
Production tubing 104
Proxy models 132, 153

R

Radial basis function neural network 262, 268
Real-time tracking 87
Recovery factor 125, 126, 129, 131–134, 136–138, 142, 145, 148, 153, 154
Reserves estimation 261
Reservoir characterisation 1, 2, 4, 57, 207–209, 213–215, 247–252
Reservoir conditions 104
Reservoir exploration 57
Reservoir management 34, 35, 261
Reservoir pressure 125, 142
Reservoirs 159, 160, 163
Residual U-net architecture 168
Rising sea levels 126
Rock physics 235, 238, 248, 250–252

S

Salt mapping 159, 163–165, 168
Salt segmentation 161, 163, 164, 178
Salt tectonics 160
Sand contamination 262
Secondary recovery 125, 126
Seismic characterisation 252
Seismic edge-detection algorithms 160
Seismic image resolution 181–184, 186, 189
Seismic imaging 181–183, 186, 201, 203
Seismic interpretation 159, 160, 162, 178, 179
Seismic salt imaging 160
Semantic segmentation 163, 169, 174
Sequential model-based optimisation 75
Shapley values 15, 16, 23, 24, 26–29
Shear sonic 57, 58, 61–63, 83
Smart well completion 104, 105, 107, 122

Smart wells 104, 105, 107–109, 114, 122
Soft computing techniques 287
Sonic waves 57, 63
Stratigraphic features 160
Structure velocity 288–291, 293–295, 297
Subsurface characterisation 160
Subsurface engineering 1, 2, 4, 5
Subsurface structures 159, 160, 164
Support Vector Machine (SVM) 7, 8, 10, 18, 208, 220–223, 225, 228, 230–233, 236, 238, 240, 243, 245, 287
Surface control and monitoring systems 105

T

Temperature 104, 105, 107, 110, 115
Tertiary oil recovery 126
Test separator 87, 91, 92
Transfer learning 159, 161–164, 168, 169, 173, 174, 177–179
Tree-based algorithms 287
Tubing head pressure 262, 269, 279, 282, 283
Tubing performance relationship 261
Two-phase flow 286–288, 291, 297, 298

U

Unconventional resources 58

V

Velocity 87, 96
Velocity model 159, 162
Virtual flow metering (VFM) 2, 87, 88, 96
Viscosity reduction 126

W

Water cuts 87, 88, 90–92
Water injection 104–108, 120, 122, 130, 132, 141, 142, 145
Water injection management 122
Water saturation prediction 247
Water-cut prediction 88
Wellhead 87, 90–93
Well-to-Seismic inversion 58
Wireline logs 35, 36, 40, 42, 43, 47, 54, 58, 60, 75, 81–83

Z

Zonal isolation 105, 107